IEE Control Engineering Series 26
Series Editors: Prof. H. Nicholson
Prof. B.H. Swanick

MEASUREMENT AND INSTRUMENTATION FOR CONTROL

Previous volumes in this series:

Volume 1	Multivariable control theory J.M. Layton
Volume 2	Lift traffic analysis, design and control G.C. Barney and S.M. dos Santos
Volume 3	Transducers in digital systems G.A. Woolvet
Volume 4	Supervisory remote control systems R.E Young
Volume 5	Structure of interconnected systems H. Nicholson
Volume 6	Power system control M.J.H. Sterling
Volume 7	Feedback and multivariable systems D.H. Owens
Volume 8	A history of control engineering, 1800-1930 S. Bennett
Volume 9	Modern approaches to control system design N. Munro (Editor)
Volume 10	Control of time delay systems J.E. Marshall
Volume 11	Biological systems, modelling and control D.A. Linkens
Volume 12	Modelling of dynamical systems-1 H. Nicholson (Editor)
Volume 13	Modelling of dynamical systems-2 H. Nicholson (Editor)
Volume 14	Optimal relay and saturating control system synthesis E.P. Ryan
Volume 15	Self-tuning and adaptive control: theory and application C.J. Harris and S.A. Billings (Editors)
Volume 16	Systems Modelling and Optimisation P. Nash
Volume 17	Control in hazardous environments R.E. Young
Volume 18	Applied control theory J.R. Leigh
Volume 19	Stepping motors: a guide to modern theory and practice P.P. Acarnley
Volume 20	Design of modern control systems D.J. Bell, P.A. Cook and N. Munro (Editors)
Volume 21	Computer control of industrial processes S. Bennett and D.A. Linkens (Editors)
Volume 22	Digital signal processing N.B. Jones (Editor)
Volume 23	Robotic Technology A. Pugh (Editor)
Volume 24	Real-time computer control S. Bennett and D.A. Linkens (Editors)
Volume 25	Nonlinear system design S.A. Billings, J.O. Gray & D.H. Owens (Editors)

MEASUREMENT AND INSTRUMENTATION FOR CONTROL

M.G. Mylroi and G. Calvert

Peter Peregrinus Ltd on behalf of the Institution of Electrical Engineers

Published by: Peter Peregrinus Ltd., London, UK.
© 1984: Peter Peregrinus Ltd.

All rights reserved. No part of this publication may be reproduced, stored in a retrieval system or transmitted in any form or by any means — electronic, mechanical, photocopying, recording or otherwise — without the prior written permission of the publisher.

While the author and the publishers believe that the information and guidance given in this work is correct, all parties must rely upon their own skill and judgment when making use of it. Neither the author nor the publishers assume any liability to anyone for any loss or damage caused by any error or omission in the work, whether such error or omission is the result of negligence or any other cause. Any and all such liability is disclaimed.

British Library Cataloguing in Publication Data

Measurement and instrumentation for control
 — (IEE control engineering series; 26)
 1. Control theory
 I. Mylroi, M.G.   II. Calvert, G.   III Series
 629.8   QA402.3

ISBN 0 86341 024 3

Printed in England by Short Run Press Ltd., Exeter

Contents

List of contributors		xi
Preface		xii

1	Electromagnetic Flowmeters	1
1.1	Introduction	1
1.2	Principle of operation	2
1.3	Velocity profile effects	6
1.4	Signal processing	8
	1.4.1 Systems with sinusoidal excitation of electromagnet	9
	1.4.2 Systems with unidirectional excitation of electromagnet	12
	1.4.3 Systems with bipolar excitation of electromagnet	14
1.5	Installation requirements	16
1.6	Factors influencing the choice of an electromagnetic flowmeter	18
1.7	Industrial applications	21
	References	23

2	Orifice Plate Flow Measurement	24
2.1	Notation	24
2.2	Introduction	25
2.3	Orifice plates and pressure tappings	27
	2.3.1 Forms of orifice plates	27
	2.3.2 Pressure tappings	30
2.4	Discharge coefficients of orifice plates	31
2.5	Installation and other effects	34
2.6	Conclusion	35
	Acknowledgements	37
	References	37

3	Other Flow Measuring Devices	38
3.1	Introduction	38
3.2	Positive displacement flowmeters	38
3.3	Turbine flowmeters	40
3.4	Ultrasonic flowmeters	41
	3.4.1 Transit time flowmeters	41
	3.4.1.1 Measurement techniques	43
	3.4.1.2 Velocity profile sensitivity	47
	3.4.1.3 Accuracy of transit time flowmeters	48
	3.4.2 Doppler flowmeters	48
	3.4.2.1 Limitations and accuracies of Doppler flowmeters	51
	3.4.3 Correlation flow measurement	51
3.5	Variable area flowmeters	52

vi Contents

	3.6	Vortex shedding flowmeters	54
		References	55

4 Two Phase Flow Measurement 61
   4.1 Introduction 61
   4.2 Historical background 61
      4.2.1 Slurry pipelines 62
      Two Phase Flow Measurement 61
      4.2.2 Pneumatic pipelines 62
      4.2.3 Pneumatic conveying 62
      4.2.4 Horizontal/Vertical conveying 62
   4.3 Measurement of Flow Velocity 63
      4.3.1 Magnetic flowmeters 63
      4.3.2 Cross-correlation flowmeters 63
      4.3.3 The cross-correlator 65
      4.3.4 Ultrasonic flowmeters 66
   4.4 Mass Flow Measurement 66
      4.4.1 Continuous belt weighing 66
      4.4.2 Capacitance flowmeters based on flow noise measurement 67
      4.4.3 Coriolis meter 68
      4.4.4 Nucleonic techniques 69
      4.4.5 Vortex and turbine flowmeters 70
   4.5 Conclusions 70
      References 70

5 Temperature Measurement 73
   5.1 The concept of temperature and the thermodynamical scale 73
   5.2 The international practical temperature scale (IPTS) 74
   5.3 Dissemination of the temperature scale 76
   5.4 Types of thermometer 76
      5.4.1 Expansion thermometers 76
      5.4.2 Thermocouple thermometers 77
      5.4.3 Resistance thermometers 81
      5.4.4 Radiation thermometry (Pyrometry) 83
   5.5 Installation and use of immersion thermometers 86
      5.5.1 Conduction errors (cold and effect) 86
      5.5.2 Self heating 87
      5.5.3 Time response 87
      5.5.4 Thermo-electric potentials 87
      5.5.5 Total or stagnation temperature 87
      5.5.6 Installation and vibration 88
      References 89
      General references 89

6 Pressure Measurement 91
   6.1 Introduction 91
   6.2 Monometer 92
   6.3 Dead weight tester 94
   6.4 Bourdon tubes, capsules and bellows 95
   6.5 Pressure transducers 97
   6.6 Capacitance type pressure transducer 97
   6.7 Reluctive type pressure transducer 98
   6.8 Force balance pressure transducer 99
   6.9 Piezoelectric pressure transducer 99
   6.10 Strain gauge pressure transducer 100
   6.11 Other pressure measuring methods and transducers 102

	6.11.1	Vibrating wire type	102
	6.11.2	Potentiometric pressure transducer	102
	6.11.3	Resistance pressure transducer	103
	6.11.4	Novel pressure transducers and pressure measurement techniques	103
		Bibliography	103

7	Force and Weight Measurement		105
	7.2	Direct comparison method	106
	7.3	Indirect comparison method	106
	7.4	Force balance technique	107
	7.5	Hydraulic and pneumatic load cells	108
	7.6	Vibrating wire force transducer	109
	7.7	Linear variable differential transformer	110
	7.8	Strain gauge force transducer	112
	7.9	Other weighing methods and force transducers	115
		7.9.1 Buoyancy principle	115
		7.9.2 Ballistic weighing	115
		7.9.3 Gyroscopic load cell	116
		7.9.4 Piezoelectric force transducer	116
		7.9.5 Magnetoelastic load cell	117
		7.9.6 Novel ideas	118
		Bibliography	118

8	Mathematical Models and their Use in Force Measuring Instruments		119
	8.1	Introduction	119
	8.2	Mathematical models and their use	120
	8.3	Application of mathematical models to force transducers	124
	8.4	Summary and conclusions	133
		References	134

9	Liquid Level Measurement		136
	9.1	Introduction	136
	9.2	Methods of level measurement	136
		9.2.1 Visual indicators	137
		9.2.2 Float actuator instruments	138
		9.2.3 Displacement type instruments	140
		9.2.4 Hydrostatic pressure instruments	142
		9.2.5 Differential pressure instruments	144
		9.2.6 Probe methods	144
		9.2.7 Radio frequency method	146
	9.3	Summary	146
		9.3.1 Visual indicators	146
		9.3.2 Float actuated instruments	146
		9.3.3 Displacement type instruments	147
		9.3.4 Hydrostatic pressure instruments	147
		9.3.5 Differential pressure instruments	147
		9.3.6 Probe methods	147
		9.3.7 Radio frequency method	147

10	Control Valves and Actuators		148
	10.1	Introduction	148
	10.2	The restricting element	149
		10.2.1 Sliding stem valves	149
		10.2.1.1 Characteristics	149
		10.2.1.2 Bonnets	150
		10.2.1.3 Size range	151

		10.2.1.4 Body forms	151
	10.2.2	Rotary Valves	152
		10.2.2.1 Ball valves	152
		10.2.2.2 Butterfly valves	153
		10.2.2.3 Size range	154
		10.2.2.4 Characteristics	154
	10.2.3	Materials of Construction	154
	10.2.4	Valve sizing	154
10.3	Actuators		155
	10.3.1	Pneumatic diaphragm actuators	155
		10.3.1.1 Valve positioners	156
	10.3.2	Piston actuators	158
	10.3.3	Electric actuators	160
	10.3.4	Actuator selection	161
		10.3.4.1 Power source	161
		10.3.4.2 Fail safe requirements	162
		10.3.4.3 Force requirements of the valve	162
		10.3.4.4 Form of control required	162
		10.3.4.5 Cost	163

11 Errors and Uncertainty in Measurements and Instruments — 164
  11.1 Introduction — 164
  11.2 Definition of Error — 164
  11.3 Measured system and measuring system errors — 164
  11.4 Errors arising in the measured system — 165
  11.5 Definition and classification of measuring system of instrumental errors — 165
  11.6 Intrinsic instrumental errors — 167
  11.7 Instrumental influence errors — 169
  11.8 Time variation of influence and intrinsic errors — 171
  11.9 Systematic and random errors — 171
  11.10 Uncertainty and its specification — 172
  11.11 Accuracy, repeatability, reproducibility — 175
  11.12 Indirect measurement — 176
  11.13 Active measurement — 177
  Bibliography — 178

12 Metrology — 179
  12.1 Introduction — 179
  12.2 Elements of measuring system — 181
    12.2.1 The variable — 181
    12.2.2 Reference — 181
    12.2.3 Calibration — 183
    12.2.4 Measurement procedure — 183
  12.3 Measurement Terms and Performance of Instrument System — 184
  12.4 Errors — 185
    12.4.1 Types of error — 185
    12.4.2 Random errors — 186
    12.4.3 Error combinations — 186
    12.4.4 Error models — 186
    12.4.5 Error propagations — 187
  12.5 Processing — 188
  12.6 Recent trends in metrology — 188
    12.6.1 The increasing use of computers — 188
    12.6.2 The increasing use of statistical techniques — 189
    12.6.3 The increasing use of coherent optics — 189

Contents ix

12.7	Conclusions	189
12.8	References	189

13 Automatic Inspection 191
  13.1 Introduction 191
  13.2 General principles 192
  13.3 The relevance of models 193
  13.4 Optical surface inspection 193
    13.4.1 Iterrogation 194
    13.4.2 Modulation 194
    13.4.3 Sensing 194
    13.4.4 Perception 197
      13.4.4.1 Automatic pattern recognition 198
      13.4.4.2 Signal detection, delineation, parameterisation 199
    13.4.5 Decision making 201
  References 203

14 Introduction to Analytical Measurement 206
  14.1 Introduction 206
  14.2 Concepts in measurement of chemical parameters 206
  14.3 Sensitivity 206
  14.4 Accuracy 207
  14.5 Precision 207
  14.6 Limits of detection 209
  14.7 Preconcentration and separation 210
  14.8 Modern techniques of analytical instrumentation 211

15 Analytical Instruments 213
  15.1 Introduction 213
  15.2 Selective spectroscopic techniques 213
  15.3 Selective electrchemical techniques 220
    15.3.1 Classification of electrochemical methods 222
    15.3.2 Potentiometry 222
    15.3.3 Polarography and voltammetry 222
  15.4 Selective chromatographic techniques 224
    15.4.1 Liquid chromatography 224
    15.4.2 Gas chromatography 229
  References 233

16 Microprocessors in Instrumentation 235
  16.1 Introduction 235
  16.2 Monitoring patients recovering from cardiac surgery 236
  16.3 Analysis of process stream composition 237
    16.3.1 Gas chromatography 238
    16.3.2 Mass spectrometry 238
  16.4 Measurement on a pneumatic conveying rig 240
    16.4.1 Overall plant description 240
    16.4.2 The air feed system 241
    16.4.3 The orifice plate calculations 241
  16.5 Correlation flow measurement 243
    16.5.1 Zero-crossing time algorithm for polarity correlation 244
    16.5.2 Implementation of the algorithm 245
      16.5.2.1 Presentation of data 245
      16.5.2.2 Functions, facilities and extras 245
      16.5.2.3 Extra functions 246
      16.5.2.4 Graphics 246

	16.5.2.5	Peak search	247
	16.5.2.6	Data checking and result checking	247
	16.5.2.7	Further result processing	248
	16.5.2.8	Peak tracking	248
16.5.3	Features for engineers		249
	16.5.3.1	Options and epROM firing	249
	16.5.3.2	Diagnostics	250
	16.5.3.3	Driving software from BASIC	250
	16.5.3.4	Communications	250
16.6 Acknowledgements			250
References			250

17 Case Study. The Development of an Instrument to Measure Coal Seam Thickness — 251

17.1 Introduction — 251
17.2 Horizontal control of cutting machines — 251
    17.2.1 The Longwall mining system — 251
    17.2.2 The need for horizontal control — 253
    17.2.3 Horizontal control of cutting machines — 254
17.3 The measurement task — 254
17.4 Possible techniques — 256
    17.4.1 The ray backscatter method — 257
    17.4.2 Sonic/ultrasonic methods — 259
    17.4.3 Electromagnetic methods — 259
    17.4.4 Mechanical methods — 260
17.5 Type 707 and 709 radiation backscatter probes — 261
    17.5.1 Type 707 probe — 261
    17.5.2 Type 709 probe — 264
17.6 The natural gamma sensor — 267
    17.6.1 Conception — 267
    17.6.2 Investigations and outline design of the natural gamma probe — 269
    17.6.3 Design of the type 801 probe — 271
17.7 Applications of the natural radiation probe — 272
17.8 Postcript - The evolution of a research and development project — 275
Acknowledgements — 277
References — 278

List of contributors

Chapter 1
E.H. Higham
Foxboro Yoxall
Redhill, Surrey RH1 2HL

Chapter 2
A.E. Spencer
Marketing Division
National Engineering Laboratory
East Kilbride, Glasgow
G75 0QU

Chapter 3
M.L. Sanderson
Electrical Engineering Dept.
University of Salford
Salford
M5 4WT

Chapter 4
C.N. Wormald
School of Control Engineering
University of Bradford
Bradford
BD7 1DP

Chapter 5
J.S. Johnston
Rosemont Engineering Co Ltd
Heath Place, Bognor Regis
West Sussex, FO22 9SH

Chapters 6 & 7
V. Erdem
Negretti Automation
Stocklake, Aylesbury
Bucks, HP20 1DR

Chapter 8
F. Abdullah
Dept. of Physics
City University
St. John Street
London EC1 4PB

Chapters 9 & 10
B. Baldwin
Education Centre
Fisher Controls Ltd
Century Works
London SE13 7LN

Chapter 11
L. Finkelstein
Dept. of Physics
City University
St. John Street
London EC1 4PB

Chapter 12
D.J. Whitehouse
Dept. of Engineering
University of Warwick
Coventry, CV4 7AL

Chapter 13
W.J. Hill
Dept. of Physics
City University
St. John Street
London EC1 4PB

Chapters 14 & 15
G.F. Kirkbright
Dept. of Instrumentation and
Analytical Science
UMIST
P.O Box 88, Manchester
M60 1QD

Chapter 16
R.M. Henry
School of Control Engineering
University of Bradford
Bradford BD7 1DF

Chapter 17
V.M. Thomas
Head of Electrical Engineering Division
NCB
Mining Research & Development Establishment
Ashby Road
Stanhope Bretby

Preface

Over the past twenty years considerable research has been undertaken on a worldwide scale into control theory and its application so that today, control principles can be applied beneficially to a wide variety of industrial situations.

However, the implementation of control strategy on plants and processes is dependent upon the availability of suitable instrumentation for sensing the relevant plant or process variables. The instruments required need to have accuracy and robustness and a level of noise immunity and here the present position is less than satisfactory. The full application of control principles in many industries is believed to be impeded by a lack of robust sensing devices.

The Science and Engineering Research Council has recognised Instrumentation and Measurement as a key area of Control Engineering and in addition to the thrust it has given to Measurement Research has included Instrumentation and Measurement in its programme of Vacation Schools. These Schools have been aimed primarily at recipients of SERC Postgraduate Studentship Awards but some non-SERC students together with a few workers from industry have also attended. It is felt important that those engaged in research in control systems, especially those who are or will be concerned with practical implementation, should have an appreciation of the problems of instrumentation and measurement. Two of these Measurement Vacation Schools have now been held, one at City University in 1979 and the second, in 1983, at Bradford University. The chapters in this book formed the lectures given at the Bradford Vacation School.

The lecture articles give an in-depth consideration of the problems of process measurement covering flow, temperature, pressure, force, weight and level, as well as control valves and actuators. Analytical measurement and analytical instruments, the problem of automatic inspection and metrology are also covered. Errors and uncertainty in measurements and instruments are discussed together with the use of mathematical modelling in sensor design and the implications to measurement of the microprocessor. An extensive Case Study concludes the series of articles and highlights the problems encountered in the development of an instrument to solve a particular industrial problem.

Chapter 1

Electromagnetic flowmeters

E. H. Higham

1.1. INTRODUCTION

The electromagnetic flowmeter was developed for measuring the flow rate of fluids in installations where the more common methods of measurement were unsatisfactory. Its principal feature is that it does not present any obstruction to the flow of fluid and provided that the liquid is an electrolyte, it is relatively insensitive to changes in fluid density, viscosity and the flow profile. It also has an essentially linear response and is particularly suitable for measuring the flow of aggressive acids and alkalis as well as slurries with coarse or fine suspended materials.

Recent general information regarding the usage or application of the various methods of flow measurement in the process industries suggests the breakdown shown in Fig.1.1.

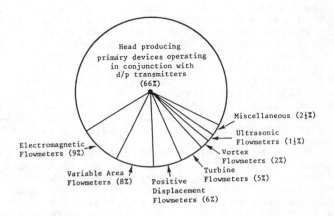

Fig. 1.1 Relative usage of the principal types of flowmeters

It should be emphasised that this information is very approximate because the method of reporting varies not only from one industry to another but also from country to country. Also, the understanding of which industries are included in the definition of process industries is vague. However the broad indications are sufficient for the present purpose.

There are many good reasons for the dominance of the orifice plate in combination with a differential pressure transmitter in the field of flow measurement. This technique has a very long history starting with the work of Bernoulli published in 1738 on which the hydraulic equations for differential pressure meters are based. In 1797, Venturi (1) described his basic work on the principles of the Venturi tube. However nearly a century was to elapse before Herschel (2) developed a commercial venturi tube for measuring large volumes of flowing fluid. The orifice plate emerged as the primary element during the first decade of this century, and since then a vast amount of operational experience and performance data has been accumulated.

On the other hand, although Faraday recognised that his law of magnetic induction applied not only to metallic conductors, but also to conductive fluids, his attempts in 1832 (3) to measure the flow of the River Thames at Westminster Bridge by observing the voltage developed between two large plate electrodes as a result of the water flowing through the vertical component of the earth's magnetic field were unsuccessful. Woolaston (4) claimed success in 1881 when he made a similar experiment using a telegraph cable that had recently been laid across the English Channel.

The first experiments to investigate the effect of a transverse field on the fluid flowing in a circular pipe were reported by Williams (5) in 1930, but apparently his interest was academic and he did not recognise the extent to which the concept could be exploited for practical measurements. It was in the medical field that the first practical designs were evolved, from 1936 onward, principally by Kolin, who identified many of the features which are to be found in modern electromagnetic flowmeters. Subsequent researchers in this field have made important contributions to the design and method of operation of these flowmeters and Wyatt (6) and his colleagues at Nuffield College, Oxford, have made very detailed studies of many of the problems which arise in applying electromagnetic flowmeters, particularly in the medical field.

The industrial use of these flowmeters appears to have originated in the Netherlands where the dredging industry needed a means for measuring the flow of sand/water slurries. The practical success of some very rudimentary equipment led to the application of the technique to other difficult flow measurements and the subsequent evolution of a commercial range of flowmeters, starting about 1955.

During the same period, several workers were studying the theoretical aspects of the flowmeter performance. For example, Thurlemann (7) determined the response of a flowmeter with a uniform magnetic field to axisymmetric flow velocity profiles, and Shercliff (8) studied various applications associated with nuclear energy, including the measurement of the flow of liquid sodium. His theoretical studies and other work are collected in his book, which still provides a standard reference on the subject.

1.2. Principle of Operation

Electromagnetic flowmeters comprise two basic parts, namely the flowtube or primary element which provides the means of transducing the flow into an emf and the transmitter which converts this emf into a signal suitable for transmission over longer distances such as a proportional dc current in the range 4 to 20 mA, a frequency in the range 0 to 10 kHz, or a pulse, in which case each pulse represents a predetermined volume of fluid.

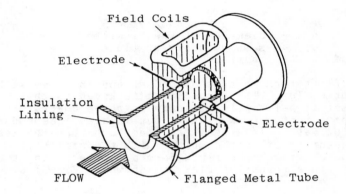

Fig. 1.2 Essential components of an electromagnetic flowmeter

The essential components of a practical electromagnetic flowtube are shown in Fig.1.2. The metal tube, fabricated from non-magnetic material such as stainless steel and fitted with flanges, provides the essential mechanical strength for the unit. It is lined with ptfe, polyurethane or other insulating material to minimise the short circuiting or diversion of the relatively small flow signal to the metal tube. The transverse magnetic field is generated by a pair of coils located on opposite sides of the tube (the core that completes the magnetic circuit is not shown), and the emf induced by movement of the fluid through the magnetic field is detected by a pair of electrodes located diametrically opposite each other with their axis perpendicular to both the magnetic field and the axes of the tube.

The electrodes are usually fabricated in non-magnetic stainless steel, although other alloys such as Monel and Hastelloy, or those based on platinum, titanium, or tantalum are sometimes used.

4 Electromagnetic flowmeters

The kinetics of the interface between the metal electrode and the fluid are not only complex but also variable and unpredictable. The only practical method of reducing the effect of these spurious signals is to measure the change in inter-electrode emf resulting from a known change in the magnetic field strength; in other words, to generate a flow measurement signal that is proportional to the flow emf divided by the magnetic field strength.

The simplest form of excitation is to use the mains supply to energise the coils of the electromagnet, in which case the flow signal is an alternating emf of the same frequency and in phase with the magnetic flux.

Thus, the induced emf is

$$e = B.d.v.\sin wt + Kw.B.\cos wt$$

where B = mean flux
 wt = angular frequency
 d = diameter of the flowtube
 v = mean velocity of fluid
and K = constant

The first term is flow dependent and in phase with the magnetic flux. The second term is known as the transformer signal and is not only in phase quadrature with respect to the flow signal, but is also independent of the flow rate. It arises from the fact that the leads to the measuring circuit combined with the effective current path through the fluid form a complete loop that is cut by the alternating flux, as shown in Fig. 1.3.

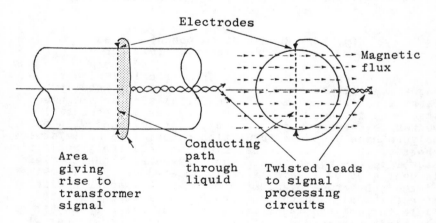

Fig. 1.3 Source of the transformer effect signal

Although careful attention to positioning the leads can reduce this signal appreciably, it is still necessary to arrange the measurement circuits so that this unwanted component is rejected. However, the phase of the flux within the tube compared with that produced by the coils and the associated laminated yoke is modified by eddy currents in the metal tube and although the effect is relatively small, it is not constant because changes of temperature change the conductivity of the metal and hence the magnitude of the eddy currents whilst variations of frequency change the mutual coupling, with the result that there is a variable zero error.

The wave forms for this type of excitation are shown in Fig.1.4, with the errors due to changes in mains frequency and temperature exagerated. For a typical flowtube 100 mm diameter the flux density would be 0.01 Tesla and the flow signal about 1 mV for a mean velocity of 1 m/s.

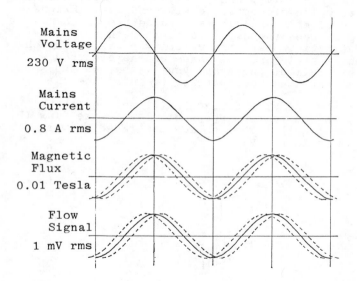

Fig. 1.4 Waveforms for an electromagnetic flowmeter with sinusoidal excitation

Referring again to the transformer signal induced by imperfect location of the signal leads, this is proportional to the projected area cut by the flux and the rate at which the flux density changes with time. If the excitation is 50 Hz, the flux 0.01 Tesla and the projected area is 3 mm^2 then this transformer signal would be approximately 10 uV. This represents 1% of the flow signal that would be developed in a typical electromagnetic flowmeter system. Since the general accuracy of these systems is usually better than 1%, the importance of restricting this error by stable mechanical contruction is apparent, although the signal conditioning circuits can ease this requirement.

One possibly unexpected aspect of flowtube design arises from the fact that the flow signal is a function of

the total flux, rather than the flux density that is cut
by the flowing fluid. Consequently, if the same measuring
circuit is to be used for a range of flowtubes, then one
of the principal problems in designing the small flowtubes
is how to concentrate the necessary total flux into the
test section, whilst for the large sizes, the problem is
that of minimising the copper loss in the coils which,
although they need only to produce the same total flux,
must, of necessity, extend over a much larger area.

1.3. Velocity Profile Effects

It is seldom possible to arrange the installation of
a flowmeter so that the flow profile is axisymmetric.
There is usually a valve, bend, or junction upstream of
the flowmeter which distorts the flow profile. Extensive
theoretical studies have been made to determine the
relationship between the signal generated in an
electromagnetic flowmeter, the velocity profile, the shape
of the magnetic field and the shape of the electrodes.
Although the detail of these analyses is beyond the scope
of this paper, it seems appropriate to introduce the
subject.

Fig.1.5 shows a wire of length (l) moving at a
velocity (V) in a direction perpendicular to the flux (B)
of a permanent magnet. The voltage generated between the
ends of the wire is B.l.V.

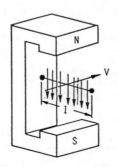

Fig. 1.5 Generation of emf in a wire
moving in a magnetic field

If the wire is now removed and a length of
non-conducting tube is placed in the magnetic field as
shown in Fig.1.6, we have the basic components of an
electromagnetic flowmeter and we may imagine filaments of
liquid spanning the tube and each generating a voltage
between their ends as they pass through the magnetic field.

Referring now to Fig.1.7, a transverse filament (a)
at the centre will generate a voltage between its ends of
B.l.V, whereas a filament (b) located nearer to the side
of the tube is shorter in length and will probably move at

Electromagnetic flowmeters 7

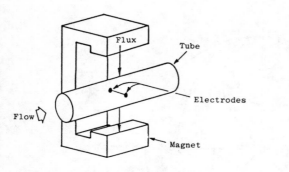

Fig. 1.6 Extension of the concept to a flowtube

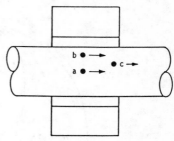

Fig. 1.7 Effect of position on induced emf

a lower velocity so that the generated voltage is, say, B.l.V/2 whilst a filament (c) nearer to the edge of the magnetic field will probably move at a velocity (V) but be in a magnetic field strength B/3 so that the voltage generated between its ends is B/3.l.V.

Thus, the potential across each filament will be different, so that if the ends of all the filaments are connected together, currents will flow in various directions and the signal developed across the filament (a) will be reduced. In spite of the complexity of these circulating currents, the actual signal (U) developed by the flowmeter for a range of ideal conditions, including a uniform magnetic field and axisymmetric flow, is

$$U = B.D.V_m$$

where B = the flux density

D = the diameter of the tube

and V_m = the mean velocity

Shercliff suggested the weight function (W) as a means for predicting the effect of distorted flow profiles such that

$$U = \int_0^r \underline{V}\, \underline{W}\, dr$$

where $\underline{V}$ = the velocity

$\underline{W}$ = the weight function

and r = the radius

Fig 1.8. shows this weight function for a uniform magnetic field. Bevir (9) has shown theoretically that

$$\underline{W} = \underline{B} \times \underline{J}$$

where $\underline{B}$ is the magnetic flux density and $\underline{J}$ the current

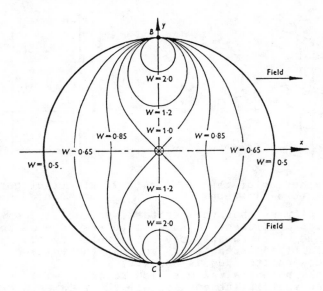

Fig 1.8 Weight function for uniform magnetic field

density that would be set up in the liquid if unit current was passed from one electrode to the other through the liquid. The requirement for the actual signal (U) developed in the flowtube to be independent of the velocity profile and only dependent on the total or mean flow rate is that $\nabla \underline{W} = 0$.

Flowtubes have been designed with weight functions much more uniform than those depicted by Shercliff for a uniform magnetic field. Consequently there is no need for such long straight upstream lengths of pipe or the inclusion of flow conditioners, as is the case for orifice plate systems, turbines, and other types of flowmeters.

1.4. Signal Processing

From a signal processing viewpoint, the flowtube can be represented as shown in Fig.1.9a. If it were possible to provide an alternating magnetic flux of stable amplitude, then measurement of the signal at the electrodes would be sufficient to provide the indication of the mean flow velocity. In practice, this is not possible; therefore the actual values of the flux and flow signals must be measured simultaneously and the flow velocity determined by dividing the flow signal by the flux signal.

The two principal methods of deriving a signal that is proportional to the flux involve measuring either the current that flows through the coils or the voltage that is applied across them (as shown in Figs.1.9b and 1.9c respectively). Both methods represent a compromise, because in neither case is the reference signal strictly proportional to the flux within the tube. The eddy currents in the metal tube as well as variations in temperature and supply frequency introduce their individual errors and it is a matter of opinion which provides a better reference. Suffice it to say that over the past decade the current reference system has become more widely used.

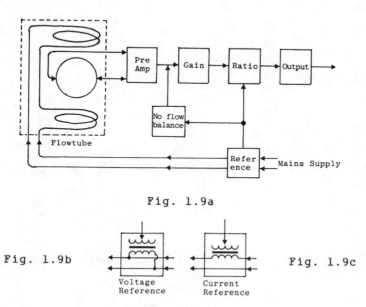

Fig. 1.9 Basic components of signal processor for ac excitation

Other methods that have been used include the provision of an additional 'search' coil in the magnetic circuit and the use of a Hall effect probe to measure the magnetic field. However both suffer from the limitation that they do not measure the field within the tube and in other respects do not show advantages which would justify the additional cost and complication.

1.4.1 Systems with sinusoidal excitation of the electromagnet.

The functional diagram of a transmitter involving this mode of operation is shown in Fig.1.10. The flowtube electrodes are connected via a special double screened cable, shown in Fig.1.11 to the balanced inputs of the high impedance input amplifier. To minimise the loading effect of the cables, the screen associated with each input signal is driven by the output from the first stage. Even so this loading does have an effect on the overall performance of the system, an effect which is determined mainly by the length of the screened cable and the conductivity of the measured fluid.

The output signals from the first stage are also applied to a differential amplifier, which provides an overall gain of about 10 but has a very high common mode rejection. The reference signal which provides the measure of the magnetic flux in the flowtube is derived from a voltage transformer connected across the supply to the flowtube.

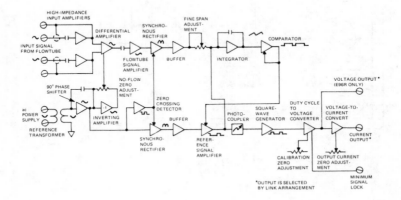

Fig 1.10 Functional diagram of transmitter

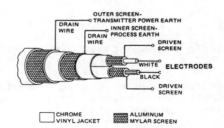

Fig. 1.11 Screened cable used for the electrode connections

The signal is first shifted in phase by 90° and then applied to the zero crossing detector to generate the synchronising signal for the two synchronous rectifiers, one of which operates on the flow signal and the other on the reference signal to generate a dc signal proportional to the magnetic flux. The same phase shifted signal is applied to an inverting amplifier with a potentiometer connected between the input and output to provide an adjustable voltage which can be combined with the output from the differential amplifier to balance out the residual no-flow signal in the system. After synchronous rectification and buffering, the two signals are applied to a circuit that produces a square wave having a duty cycle linearly proportional to the flow signal.

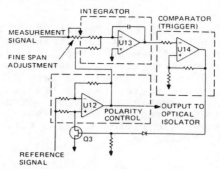

Fig. 1.12 Signal dividing stage in the transmitter

This comprises an integrator, a comparator and a reference signal amplifier whose gain is switched between +1 and -1 as shown in Fig.1.12. In operation, when the flow signal is zero, the reference voltage alone drives the integrator until its output reaches the trigger level of the comparator, whereupon the polarity of the output from the reference amplifier is reversed; as a result the integrator output is driven in the opposite direction until the comparator resets. Under these conditions the comparator output is a square wave having equal mark/space ratio (or unity duty cycle) as shown in Fig.1.13. When a flow signal is applied, the integrator ramps more quickly in one direction and more slowly in the other causing the mark/space ratio of the comparator to vary in proportion as shown in Fig.1.14.

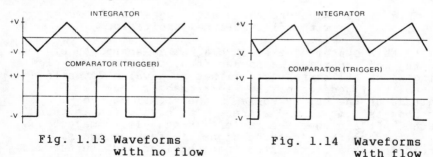

Fig. 1.13 Waveforms with no flow

Fig. 1.14 Waveforms with flow

This pulse drives a photo-coupler to provide the necessary galvanic isolation between the input and output circuits. Thereafter the signal is shaped and applied to a 'mark/space' ratio-to-voltage converter which generates a proportional signal in the range 0 to 10 V dc, which may be used as input to a variety of controllers, alarms and signal conditioners. The same signal is also used within the transmitter to operate a moving coil meter if this feature is specified and to drive a voltage-to-current converter which generates the standard 4 to 20 mA dc (or 10 to 50 mA dc) signal for transmission purposes.

1.4.2 Systems with unidirectional excitation of electromagnet.

Systems of the general type described in the previous section have been in service for more than two decades during which numerous refinements have been introduced in both the mechanical construction and the measurement circuits. But the need to improve the zero stability and to reduce the power consumption led to the introduction of alternative modes of operation. An improvement in both respects was effected by replacing the sinusoidal excitation of the magnetic field at mains frequency with a system shown in Fig.1.15.

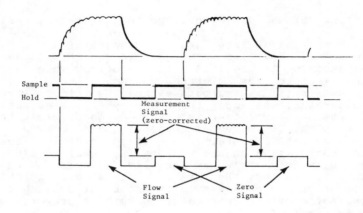

Fig. 1.15 Waveform for electromagnetic flowmeter with unidirectional excitation

In this system, the electromagnet is energised by applying the rectified mains and then, after a few cycles during which the current stabilises at a value determined by both the impedance of the electromagnet and the magnitude of the supply voltage, the signal at the electrode is sampled and stored. The excitation is then switched off and when the current in the electromagnet has fallen to zero the electrode signal is sampled again. Subtracting this signal from that obtained and stored when the electromagnet is energised provides a measure of the flow rate. In this way, errors due to the variable kinetics of the electrode/liquid interface are greatly reduced and a zero adjustment is not required.

The flowtubes required for this mode of operation are virtually the same as those used for the ac excited systems. However, since the measurement signals are sampled when the flux is steady except for the small ripple at twice mains frequency, the transformer effect signal is virtually eliminated and the magnetic flux is more closely related to the coil current because the eddy

currents which are induced by the alternating magnetic
field have fallen to a relatively small value by the time
the flow signal is sampled and the magnetic field which
they create has a negligible effect on the main magnetic
field. Fig.1.16 shows the functional diagram of the
transmitter.

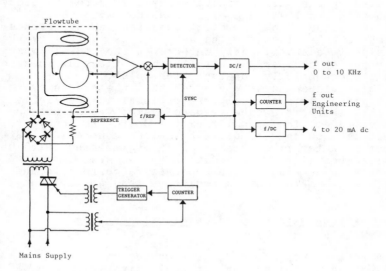

Fig. 1.16   Functional diagram of transmitter

A signal from the mains is applied to a counter which
provides two output signals, one of which synchronizes the
trigger generator and the other, the detector circuit.
The signal to the trigger generator makes it operative for
eight cycles of the mains supply, after which it is held
inoperative for the next eight cycles of the mains and the
sequence is then repeated. As a result, the transformer
is energized from the mains supply causing a direct
current with a mains frequency ripple superimposed on it
to build up in the flowtube coils. The magnetic field
which this creates causes a small voltage, proportional to
the flow rate of liquid through the flowtube to be
developed across the electrodes. After amplification, the
signal is applied as one of the inputs to a summing
junction.
Referring again to the counter, its second output is
generated, starting at the beginning of the fourth mains
cycle to the end of the eight and again from the twelfth
through to the sixteenth cycle. It is used to synchronize
the detector, (which is in effect an integrator) so that
it is only operative during the period when the magnetic
flux is either zero or has stabilised at the high value
but with the small superimposed ripple.

The detector output is applied to a dc-to-frequency stage which generates a pulse of constant duration and at a repetition rate proportional to the dc input. During the period of the second output from the counter, the signal developed across a resistor, connected in series with the coils, is used to generate a reference voltage that is proportional to the magnetic flux. In the frequency/reference circuit, this reference voltage is modulated by the pulses generated by the dc-to-frequency stage. The resultant signal is combined with the amplified signal from the electrodes and the difference is applied to the detector. The detector is 'gated' by the sync signal so that it only integrates the signals which appear during the periods when the magnetic flux is zero or at its essentially constant high value.

This feedback system ensures that the frequency output signal is accurately proportional to the liquid flow rate in the flowtube, regardless of variations in mains voltage. The signal from the dc-to-frequency circuit is also used to drive a scaling counter so that the output which it generates is in engineering units such as pulses per litre. In addition there is a frequency-to-dc stage which generates a current in the range 4 to 20 mA dc into a resistive load up to about 1500 ohms.

1.4.3 Systems with bi-polar excitation of the electromagnet.

A further refinement of the concept involves energisation of the magnetic circuit first with one polarity for a fixed period, then removing the excitation for a similar period before energising it in the reverse direction, and so on. The advantage claimed for this system is that it further reduces the errors due to polarisation at the electrode fluid interface. Also, the coil drive circuits are arranged to optimise the speed at which the current reaches a stable value at which it is held during the signal sampling process.

Thus, the speed of response is enhanced and the effect of eddy currents is eliminated because these only have an effect whilst the current is changing and the measurement signal is taken after they have died away. Also the transformer signal is virtually eliminated.

Fig.1.17 shows how the excitation of the magnetic circuit is synchronised with the mains supply. The coil current is switched on for a period of four cycles of the mains after which it is switched off for four cycles. After this quiescent period the coil current is switched on again for four cycles but the current now flows in the reverse direction. At the end of this period it is switched off for four cycles and then the entire sequence is repeated.

During each period in which current is applied to the coil, the logic circuits synchronise the sampling and holding for both the flow signal and the coil current, as well as the operation of the voltage ratio to duty cycle conversion. This latter circuit operates at a nominal 1 kHz with the duty cycle arranged to be 15% for the lower range value and 85% for the upper range value.

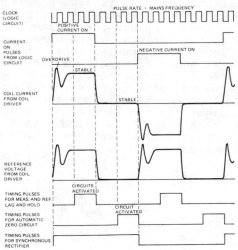

Fig. 1.17   Waveforms for an electromagnetic flowmeter with bipolar excitation

Fig.1.18 shows how other optional features are provided. These include an integral moving coil meter to monitor the output signal, a digital flow rate meter and pulse output for driving an electromechanical counter. This circuit is adjustable between 0.1 and 10 Hz (corresponding to the upper range value) and provides a 24 V signal 50 ms duration at 0.25 A. In other respects, the instrument has evolved from previous designs.

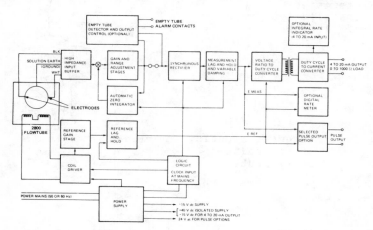

Fig. 1.18   Functional diagram of the transmitter

1.5. Installation Requirements.

The flowtubes themselves may be mounted in any position provided that the associated pipework and plant are arranged so that the flowtube is always full. The preferred arrangement is for the axis of the flowtube to be vertical and the flow to be upwards as shown in Fig.1.19. This is particularly so if the process fluid is a slurry. If the flowtube is mounted other than with its axis vertical then it should be orientated so that the axis of the electrodes is horizontal, as shown in Fig. 1.20, to minimise the effect of entrained bubbles which not only give rise to volumetric errors but may also interrupt the flow signal to the electrodes. The transmitters are available for integral mounting on the flowtube as shown in Fig.1.21 or separately on a 50 mm (2 inch) pipe or on a surface.

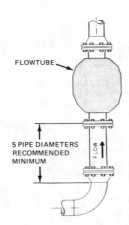

Fig. 1.19 Preferred installation of flowtube

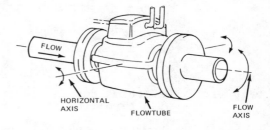

Fig. 1.20 Orientation of electrodes with flowtube horizontal

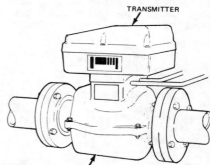

Fig. 1.21 Integral mounting of transmitter on flowtube

For the systems in which the flowtube is excited from the mains supply, it is important to arrange the wiring so that the mains supply is taken to the flowtube via the transmitter rather than from a separate circuit, as load changes in one or other of the two circuits could cause phase shifts of one supply with respect to the other and this in turn would modify the synchronisation of the phase sensitive detectors in the transmitter and so give rise to measurement errors.

As mentioned previously, it is desirable to arrange an installation so that the velocity profile of flow is well developed. This was particularly true for earlier designs of flowtube, but for present designs a straight section 5D long immediately upstream of the flowtube provides sufficient conditioning to reduce velocity profile errors to less than 1%. Electrical continuity between the flowing liquid and the metal body of the flowtube is required to provide the reference potential for the measurement signal. With unlined metal pipes connected to the flowtube, continuity is provided via the flange bolts, but with lined or non-conducting pipes, earthing rings must be fitted at each flowtube flange as shown in Fig.1.22.

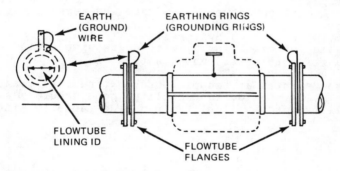

Fig. 1.22 Installation of earthing rings

These rings are circular metal plates each having a tab for the electrical connection and a concentric hole slightly smaller than the bore of the flowtube. In some modern flowtubes, the earthing rings are included in the basic design to minimise the calibration errors which would otherwise be caused by distortion of both the magnetic and electric field within the flowtube due to the adjacent flanges and pipework being fabricated from, magnetic material, non-magnetic but conducting material, or non-magnetic and non-conducting material.

1.6. Factors Influencing Choice Of An Electromagnetic Flowmeter

As has been mentioned previously, the majority of flow measurements in the process industries are made using a primary device (such as an orifice plate) to create a head loss and hence a differential pressure that can be measured by one of the many different types of transmitter that are now available. Fig.1.23 shows that, for line sizes greater than about 75 mm, this is the least expensive system and it is supported by a vast amount of practical experience. The measurement or output signal can be a pneumatic pressure, electronic current or frequency, and there are many suppliers of these devices. Consequently choice of an electromagnetic flowmeter system must be justified by one or more of the following features;

1. Low head loss, and hence suitable for measuring the flow of slurries.
2. Insensitive to changes in temperature, pressure, density and viscosity of the process fluid.
3. Non-invasive sensors and hence suitable for applications where strict requirements for hygiene are imposed.
4. Linear relation between flow rate and measurement signal.
5. No moving parts, no hysteresis.
6. Suitable for measuring flow of aggressive fluids.
7. Insensitive to swirl and pulsatile flow.

These features are supplemented by the more general characteristics of wide rangeability, good accuracy and repeatability, and availablity in a wide range of sizes.

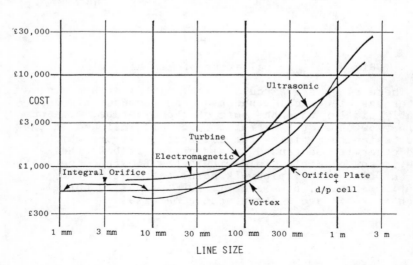

Fig. 1.23   Costs versus line size for various types of flowmeters

However, there are two distinct limitations, namely:

1. The fluid must be an electrolyte (i.e. conductive) and consequently the system is unsuitable for measuring the flow of gases or liquid hydrocarbons.
2. The systems require relatively high power for their operation and therefore they cannot be intrinsically safe.

In spite of this, the unique features of electromagnetic flowmeters are sufficient to sustain their position as a principal alternative to the orifice plate/differential pressure systems in spite of the higher cost.

Figs.1.24 and 1.25 provide a comparison with the other methods of measurement.

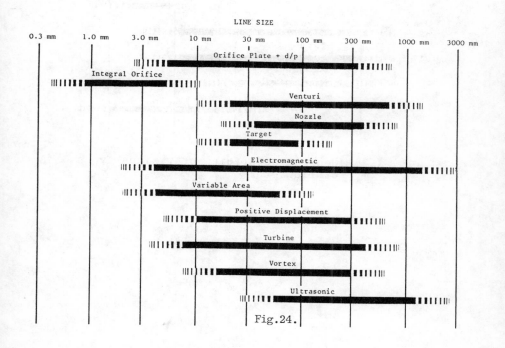

Fig. 1.24 Availability of flowmeters for various line sizes

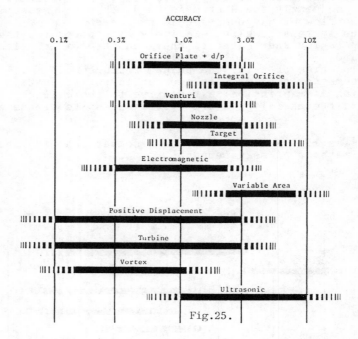

Fig. 1.25 General accuracies of various types of flowmeters

1.7. Industrial Applications.

There are so many applications of electromagnetic flowmeters involving only a single instrument and its associated control loop that to cite them would do little more than illustrate how one or more of the unique characteristics of this type of flowmeter is used to overcome what otherwise would be a difficult flow measurement problem.

An example of several electromagnetic flowmeters operating in a coordinated system is the blending of the various types of paper stock and additives for a paper making machine.

Fig.1.26 shows how the flowmeters are arranged to measure the flow of softwood, hardwood and broke stock so that it can be blended continuously under control of a digital blend control and the correct quantities of dye and additives such as clay, alum and size introduced in the required proportions. The paper stock, which typically includes between 3 and 5% by weight of solid material, is a particularly difficult material to meter, and the non-invasive sensors, together with the high accuracy that can be achieved, make the electromagnetic flowmeter the only instrument suitable for this

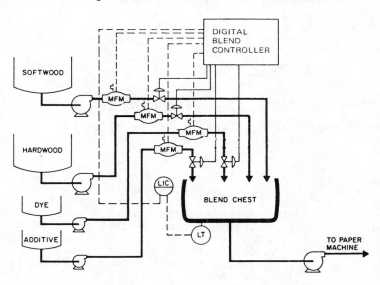

Fig. 1.26 Use of electromagnetic flowmeters for paper stock blending

application. The dye, alum, clay and size are also difficult materials to meter but can be handled by electromagnetic flowmeters. In an actual application, the transmitters would be provided with the pulse output circuits so that a digital blending controller operating in conjunction with a level controller maintains the correct level in the blend chest for supplying to the paper making machine. Another example of the use of several electromagnetic flowmeters in a process is the SO_2 scrubber shown in Fig.1.27.

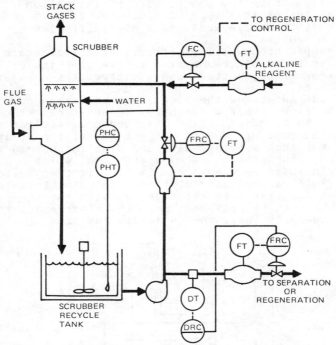

Fig. 1.27 Use of electromagnetic flowmeters in an SO^2 scrubber

The scrubbing action is effected by spraying a lime based slurry into the flue gases as they rise up the scrubber tower. The liquid that collects in the bottom of the tower is transferred to the recycle tank where the pH is measured and any deviation from the desired value is transmitted as the setpoint to the flow controller, whose measurement signal is derived from the electromagnetic flowmeter in the lime slurry reagent line. In this way, the pH is controlled accurately to prevent underabsorption of SO_2 at low pH values which also leads to scaling and plugging of the scrubber nozzles as well as unnecessary usage of the lime slurry reagent. A further control loop includes a density transmitter which in conjunction with another electromagnetic flowmeter controls the rate at which the process fluid is withdrawn for separation or regeneration. The fluids in this process all have entrained solids so that the elctromagnetic flowmeter provides the only reliable method for measuring their flow rates.

REFERENCES

1. Venturi, G.B., Researches experimentale sur le principle de la commumication lateral du mouvement dans les fluides applique a l'explication de differentes phenomenes hydrauliques.

2. Herschel, C., The Venturi water meter Trans. ASCE 1887.

3. Faraday. M., Experimental researches in electricity. Phil Trans. 1832, 15, 175.

4. Woolaston, C., Tidally induced e,m,f,s in cables. J. Soc. Tel. Engrs., 1881, 10, 50.

5. Williams, E.J., The induction of e,m,f,s in moving liquid by a magnetic field and its application to an investigation of the flow of liquids. Proc. Phip. Soc. London 1930, 42, 466.

6. Wyatt, D.C., Problems in the measurement of blood flow by magnetic induction. Phys. Med. Biol. 1961, 5. 289 + 369.

7. Thurlemann, B., On the electric speed measurement of fluid. Helv. Phys. Acta 1941, 14, 383.
   Method of electric speed measurement of fluid. Helv. Phys. Acta. 1955, 28, 483.

8. Shercliff, J.A., The theory of electromagnetic flow measurement C.U.P. London 1962.

9. Bevir, M.K., The theory of induced voltage electromagnetic flow meter. J. Fluid Mech. 1970, 43, 577-590.

Chapter 2

Orifice plate flow measurement

Dr. A. E. Spencer, OBE

2.1 NOTATION

A_1	Area of upstream pipe
A_2	Area of throat or orifice
C	Discharge coefficient of flowmeter
D	Upstream diameter of flowmeter or inlet pipe
d	Throat or orifice diameter
m	Area ratio, $\{= (d/D)^2\}$
P_1	Absolute pressure upstream of flowmeter
P_2	Absolute pressure at throat or downstream tapping
Δp	Pressure difference $(= P_1 - P_2)$
Q	Flowrate
r	Pressure ratio, ie $\left(1 - \dfrac{P_1 - P_2}{P_1}\right)$
$\bar{v}_1$	Mean velocity at upstream section
$\bar{v}_2$	Mean velocity at throat or downstream section
α	Flow coefficient
β	Flowmeter diameter ratio, $(= d/D)$
γ	Ratio of specific heat, (C_p/C_v)
ε	Expansibility factor
ρ	Density of flowing fluid at upstream section
μ	Dynamic viscosity of the fluid
ν	Kinematic viscosity of the fluid $(= \mu/\rho)$

2.2 INTRODUCTION

The orifice plate flowmeter is one of a group of flowmeters which operate by creating a difference in static pressure between the upstream and downstream side of the device. It may be installed either 'in-line', ie with pipework both upstream and downstream of it, or at the inlet to or outlet from a length of pipe. Other types of differential pressure meters are:

nozzles   low-loss devices (eg Dall tube)
venturi tubes   inlet flowmeters
venturi nozzles   variable area meters
                       drag plate flowmeters (Target meter).

All together it is estimated that well over 60 per cent of the flowmeters in use in industry at the present time are differential pressure devices, with the orifice plate being the most popular in this country. The nozzle has been in favour especially for steam flow measurement in Germany while venturi tubes were used almost inclusively for measurements in large water mains around the world until the advent of the electromagnetic flowmeter.

The main advantages of pressure difference flowmeters are:

a   they are simple to make, containing no moving parts,

b   their performance is well understood,

c   they are cheap - especially in larger pipes when compared with other meters,

d   they can be used in any orientation, and

e   they can be used for most gases and liquids.

Their main disadvantages are:

i   their rangeability (turndown) is less than for most other types of flowmeter,

ii   significant pressure loss may occur,

iii output signal is non-linear with flow,

iv   the discharge coefficient and the accuracy may be affected by the pipe layout and/or the nature of flow, and

v   they may suffer from ageing effects, ie the build-up of deposits or erosion of sharp edges.

The fundamental Bernoulli equation which deals with the relationship of the static and kinetic energy along

26 Orifice plate flow measurement

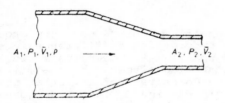

Fig.2.1 Converging flow

streamlines in a fluid stream can be used. Simplified, this shows that in a converging pipe, Fig. 2.1, the energy equation between two planes is:

$$P_1 + \tfrac{1}{2}\rho \bar{v}_1^2 = P_2 + \tfrac{1}{2}\rho \bar{v}_2^2. \qquad (2.1)$$

The other equation which can be used is the continuity equation which states that the mass flowing past any cross-section in a pipe remains constant. For incompressible flow this can be written as:

$$Q = A_1 \bar{v}_1 = A_2 \bar{v}_2 \qquad (2.2)$$

where

$$\frac{A_2}{A_1} = m \quad \text{(area ratio)}. \qquad (2.3)$$

By substituting for $\bar{v}_1$ and $\bar{v}_2$ it can be seen that

$$Q = \frac{A_2}{\sqrt{(1 - m^2)}} \sqrt{\left\{\frac{2(P_1 - P_2)}{\rho}\right\}}. \qquad (2.4)$$

This equation is valid only if no losses occur and if the moving fluid completely fills the areas A_1 and A_2.

In reality neither of these assumptions is valid and the discharge coefficient, C, has to be introduced to compensate. The discharge coefficient can therefore be defined by,

$$C = \frac{Q \sqrt{(1 - m^2)}}{A_2 \sqrt{\{2(P_1 - P_2)/\rho\}}}. \qquad (2.5)$$

The term $\sqrt{(1 - m^2)}$ is known as the velocity of approach factor. Where $A_1 \gg A_2$ and therefore $\bar{v}_1 \ll \bar{v}_2$, $\sqrt{(1 - m^2)}$ tends to unity. Instead of the area ratio the preferred term used internationally is the diameter ratio, β, where:

$$\beta = \frac{d}{D} \quad \text{and} \quad m = \beta^2 = \frac{d^2}{D^2},$$

the diameter d being the bore diameter and D being the upstream pipe diameter.

Another form of coefficient, the 'Flow Coefficient' α, has been commonly used in Europe and this is given by

$$\alpha = \frac{Q}{A_2 \sqrt{\{2(P_1 - P_2)/\rho\}}} \qquad (2.6)$$

In nozzles and venturi tubes the flow follows the boundary of the walls closely and the values of C are usually close to unity. However, in the case of the orifice plate the flow continues to converge downstream of the plate forming a 'vena contracta'. The area of this cannot practically be measured and is thus not known accurately. In calculating the discharge coefficient therefore, the area at the orifice bore is used which leads to a value of C of approximately 0.6. This, in effect, includes a coefficient of contraction.

If the fluid being metered is compressible, there will be a change in density when the pressure of the fluid falls from P_1 to P_2 on passing through the device. As the pressure changes quickly, it is assumed that no heat transfer occurs and because no work is done by or on the fluid, the expansion is isentropic. In nozzles and venturi tubes the expansion is almost entirely longitudinal and an expansibility factor, ε, can be calculated assuming one-dimensional flow of an ideal gas. For orifice plates the expansibility correction factor has to be determined experimentally both because the contraction is not known exactly and changes occur in the jet. The expansibility factor is thus a function of the diameter ratio, the specific heats and the pressure ratios (P_2/P_1).

The full equation for the orifice plate flowmeter in compressible flow is therefore

$$Q = \frac{C}{(1 - \beta^4)^{\frac{1}{2}}} \varepsilon \frac{\pi}{4} d^2 \sqrt{\left\{\frac{2\Delta p}{\rho}\right\}}. \qquad (2.7)$$

2.3 ORIFICE PLATES AND PRESSURE TAPPINGS

2.3.1 Forms of Orifice Plates

Many different geometrical profiles have been tried in order to obtain constant discharge coefficients over as wide a range of flowrates as possible. Examples of these can be found in the literature but the distinctive features of these devices are that there is a flat front and back face and the variations are in the profile of the bore and to the immediately adjacent areas of the face on either side.

28 Orifice plate flow measurement

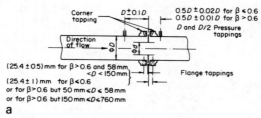

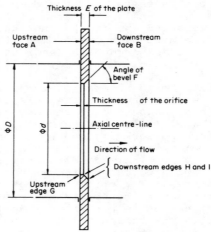

Fig.2.2 (a) Location of ISO 5167 pressure tappings upstream and downstream of orifice plate
(b) Details of ISO 5167 orifice plate

From the early days of the orifice plate in the 1890s however the most widely used plate has been one in which the bore diameter is cut out to give a square edge with the upstream face, Fig. 2.2. At the downstream end, the outlet may be bevelled if for strength reasons the thickness of the plate exceeds 0.02D but in all cases the maximum thickness of the plate is limited to 0.05D. This is because the streamlines of the emerging jet will be influenced and hence the contraction coefficient and the disharge coefficient will be affected if these limits are exceeded. At the other extreme a minimum of 0.005D has been adopted for the overall thickness of the plate.

This form of orifice plate, commonly known as a 'sharp-edge' orifice plate, was standardised in the USA in the 1920s, followed shortly afterwards by standardisation by DIN in Germany and then adopted in the late 1930s by the International Standards Association. ISA subsequently became the International Organisation for Standardization, ISO, under whose aegis the world's flow measurement standards have subsequently been published.

The present version of the international standard on orifice plates is ISO 5167 (1). The present author has described in Reference 2 how this standard has emerged from the earlier work and the dominant features which are at present the subject of some controversy and the objective of major experimental programmes in this and other countries in Western Europe and in the USA and Japan.

While the normal profile used in orifice plates is thus square or sharp edged, conical-entrance and quarter-circle plates are also used, especially for viscous flow. The downstream edge of a square-edged plate can be bevelled, unless the plate is thin, whereas the downstream edges of conical-entrance and quarter-circle plates are square. On the other hand the upstream edge is a cone or circle and these shapes show a near-constant coefficient down to quite small Reynolds numbers (3). For special purposes, eccentric or even non-circular orifice plates are used; for example, in metering suspended solids a chord-type orifice plate can be used, Fig. 2.3.

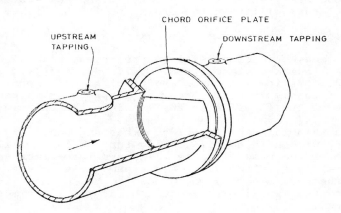

Fig.2.3 Chord-type orifice plate

Having established the general concept of a constant discharge coefficient it will seem strange to refer to limits on the flow range for which orifice plates can be used. In practice, however, fluids are not ideal and frictionless and the velocity distribution in a pipeline as well as the turbulence pattern, even if the pipeline is very long and uniformly straight, will change with the characteristics of the fluid and its flowrate through the pipe.

A parameter which has been found to give a generalised picture of the flow pattern inside such a pipeline is called the Reynolds number. It is defined by:

$$Re = \frac{\rho \bar{v} d}{\mu} = \frac{\bar{v} d}{\nu}. \qquad (2.8)$$

The mean velocity in the pipeline, $\bar{v}$, is clearly a simplified idea of what is really happening, just as the fluid density has to be an average value, but in the case of a gas may well vary across the cross-section.

To a first approximation, however, the discharge coefficient of an orifice plate will bear a direct relationship to the upstream Reynolds number which is for any specific flowing fluid directly related to the flowrate so that:

$$C = f(Re) = f(Q). \qquad (2.9)$$

It will be appreciated that this means that the same discharge coefficient value will be obtained with different combinations of fluid and flowrate.

Taking water, with a kinematic viscosity of 1 cSt as a base, a Reynolds number of 100 000 would be obtained in a 100 mm pipeline for a mean fluid velocity of 1 m/s. It is predicted that the discharge coefficient for an orifice plate of 0.5 diameter ratio with corner tappings (to be described later) will then, according to ISO 5167, be:

$$C = 0.6053.$$

The same coefficient will be predicted, however, if a gas with a kinematic viscosity of 0.015 cSt is flowing through this orifice meter-run and pipeline at a mean velocity of nearly 15 m/s. The mass flowrate will then be 1/100 that of the flowrate of the water in the example above.

To obtain the same Reynolds number and hence the same coefficient with a thick oil flowing through the pipeline and flowmeter the mass flowrate will have to be increased to get a higher mean velocity. For a kinematic viscosity of 200 cSt, the mean velocity would need to be an impossible 200 m/s and the mass flowrate would be around 150 times that of the water.

This flow measuring device is not suitable for Reynolds numbers approaching the transition and laminar flow zones because the discharge coefficient rises and then falls rapidly in a way which is difficult to predict with any accuracy. It comes into its own with increasing Reynolds numbers in the turbulent region. Indeed tests up to about 40 million suggest that the changes to the discharge coefficients for all Reynolds numbers above 1 million do not exceed about 0.25 per cent. This is applicable to long straight pipes and what are termed fully-developed flow conditions.

2.3.2 Pressure Tappings

The previous section has dealt with variations in the geometrical profile of the orifice plate and emphasised that the most common form is that standardised in ISO 5167 as the 'sharp edged' orifice plate.

In the early days the pressure difference across the orifice plate was measured from any pair of tapping holes drilled in the pipe upstream and downstream of the orifice plate. In the early 1900s different companies adopted their own preferred locations and claimed various advantages for these. In the USA the flange taps won the day and became standardised in ASME Fluid Meters. These are drilled through the flanges perpendicular to the pipe at distances of 1 inch upstream and 1 inch downstream from the faces of the flanges.

In Germany corner tappings were preferred and standardised by the VDI and DIN, the equivalents to the Institution of Mechanical Engineers and the British Standards Institution in the United Kingdom. These tappings are drilled so that they are at a slight angle so that they come out with one edge of the tapping hole just touching the face on either side of the plate. These tappings were also generally adopted by other European countries.

In the USA vena contracta taps were accepted as an alternative for many years but are inconvenient if the orifice plate bore has to be changed to adapt the pressure difference obtained with changing flowrates. The firm of George Kent's in the UK and others had found D and D/2 tappings just as good as vena contracta taps and since they have the advantages of being fixed and non-dimensional they have been accepted in ISO 5167.

Thus there are three standard locations for the differential pressure tappings to be placed in the pipe, see Fig. 2.4.

a   corner tappings,

b   flange tappings, and

c   D and D/2 tappings.

2.4 DISCHARGE COEFFICIENTS OF ORIFICE PLATES

J Stolz, Chairman of the ISO Technical Sub-committee responsible for drafting ISO 5167, led the way to a new understanding of the coefficient relationship with its many controlling parameters. In the late 1960s and early 1970s the attempts to co-ordinate the various data bases built up over the preceding decades had been principally mathematical. It will be appreciated that early experiments had then to be discarded because they were based on non-standard devices but there still remained many problems in trying to develop internationally acceptable coefficient predictions. Often the raw data were not available so that curves arbitrarily fitted to smoothed data had to be interpreted to provide a suitable data base.

Equations using power series were tried and while they may have been excellent fits with the data used to obtain them they were inconsistent with each other, often difficult to use, and dangerously wrong if extrapolated. For example one such equation had terms in powers of the diameter ratio going up to β^{24}.

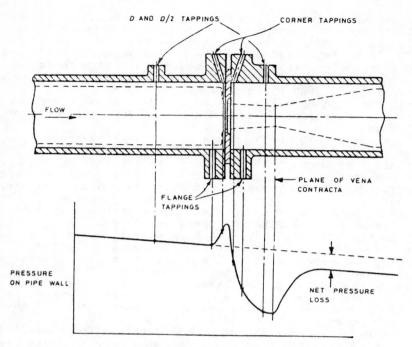

Fig. 2.4 Alternative pressure tappings

Stolz realized that improvement could only come by recognising that the discharge coefficients derived from different sets of pressure tappings must nevertheless be related to each other by physical laws. Thus the results for flange tappings and corner tappings must become identical for large pipe sizes since the allowable tolerances on their physical locations then overlap. Similarly flange and D and D/2 tappings must give the same results at very small sizes. Again all coefficients for all tappings and pipe diameters must approach each other as the diameter ratio decreases. An equation was then fitted to the non-dimensional measurements of the pressure distribution just upstream and downstream of the orifice plate based on independent tests. Combining these boundary conditions to represent logically these laws with the two sets of data which were the best authenticated (Beitler's in USA and Witte's in Germany), he evolved an equation for calculating the discharge coefficient which was relatively straightforward.

The Stolz equation (Tables 2.1 and 2.2) is not necessarily valid for all time but can, with great confidence, be regarded as a satisfactory foundation for accommodating new data merely by altering the constants. In the author's opinion any updated correlations of data should be based on its principles.

Orifice plate flow measurement 33

Table 1 The Stolz equation*

$$C = 0.5959 + 0.0312\beta^{2.1} - 0.184\beta^8$$
$$+ 0.0029\beta^{2.5}\left[\frac{10^6}{Re_D}\right]^{0.75}$$
$$+ 0.0900L_1\beta^4(1-\beta^4)^{-1} - 0.0337L'_2\beta^3$$

If $L_1 \geqslant \dfrac{0.0390}{0.0900} (= 0.4333)$

use 0.0390 for the coefficient of $\beta^4(1-\beta^4)^{-1}$

* As given in ISO 5167 (clause 7.3.2.1)

Table 2 Values of L_1 and L'_2

Corner tappings	$L_1 = L'_2 = 0$
D and $D/2$ tappings	$L_1 = 1^2$; $L'_2 = 0.47$†
Flange tappings	$L_1 = L'_2 = 25.4/D$‡

† Hence coefficient of $\beta^4(1-\beta^4)^{-1}$ is 0.0390
‡ Where D is expressed in mm

The coefficients given by the Stolz equation are only applicable to the type of orifice plates specified in ISO 5167 (Fig. 2.1) and can only be applied when the conditions of use given in Table 2.3 are met (ISO 5167, clause 7.3.1). Another condition is that the upstream pipeline shall be smooth with an upper limit of relative roughness generally less than k = 0.001D (this is approximately equivalent to the surface obtained in a 50 mm (2 in) diameter new seamless cold drawn steel pipe). The limiting value of k is however dependent on the diameter ratio of the plate being used.

Table 3 Conditions of validity

	Corner taps	Flange taps	D and $D/2$ taps
d (mm)	$\geqslant 12.5$	$\geqslant 12.5$	$\geqslant 12.5$
D (mm)	$50 \leqslant D \leqslant 1000$	$50 \leqslant D \leqslant 760$	$50 \leqslant D \leqslant 760$
β	$0.23 \leqslant \beta \leqslant 0.80$	$0.2 \leqslant \beta \leqslant 0.75$	$0.2 \leqslant \beta \leqslant 0.75$
Re_D	$5000 \leqslant Re_D$ $\leqslant 10^8$ for 0.23 $\leqslant \beta \leqslant 0.45$ $10\,000 \leqslant Re_D$ $\leqslant 10^8$ for 0.45 $< \beta \leqslant 0.77$ $20\,000 \leqslant Re_D$ $\leqslant 10^8$ for 0.77 $\leqslant \beta \leqslant 0.80$	$\geqslant 1260\beta^2 D$† $\leqslant 10^8$	$\geqslant 1260\beta^2 D$† $\leqslant 10^8$

† Where D is expressed in mm

Inevitably there will be uncertainties both in the physical measurements to be entered into the flow equation (equation (2.7)) and those associated with the discharge coefficient equation (Table 1). All these sources of uncertainty must be taken into consideration when assessing the overall accuracy of a measurement and another ISO standard, ISO 5168 (4), provides guidance on how to determine this overall uncertainty. The user must estimate the uncertainties associated with his own measurements, but those associated with the discharge coefficient required when dealing with gases particularly at pressures near to ambient, have to be specified. The uncertainties published in ISO 5167 associated with the Stolz equation are given in Table 2.4.

Table 4 Uncertainty associated with the Stolz equation*

When β, D, Re_D and k/D are assumed to be known without error, the uncertainty of the value of C is:

	Corner taps	Flange taps	D and $D/2$ taps
$\beta \leq 0.6$	0.6%	0.6%	0.6%
$0.6 \leq \beta < 0.8$	β%	—	—
$0.6 \leq \beta \leq 0.75$	—	β%	β%

* *ISO 5167 Clause 7.3.3.1*

As mentioned earlier, different tapping arrangements evolved by custom in different places standardising on such locations out of the infinite number of possibilities which could have been chosen. Many have been the discussions and arguments in favour of these and other arrangements and the Stolz equation gives, theoretically, the coefficient which might be expected in any combination. The best trusted presently available data, however, are based on the three tapping arrangements illustrated in Fig. 2.1 and referred to in Tables 2.2-2.4.

2.5 INSTALLATION AND OTHER EFFECTS

The international standard ISO 5167 lays down quite stringent conditions for the manufacture, installation and use of orifice plates. If an accuracy of 3 per cent or better is considered important then it is vital to conform to these specifications. If 5 per cent is adequate it is still necessary to comply with most of the requirements but provided the Reynolds number is sufficient, then a simple coefficient value of 0.61 could be adopted for C in equation (2.7) given earlier. It has to be remembered that the expansibility correction factor will still need to be applied, however, unless the pressure ratio is above 0.975.

If accuracies of 2 per cent and better are sought then one of the main considerations must be to ensure that the flow pattern into the orifice meter run is reasonable. All too often the meter is installed just downstream of a series of bends or with a control valve located a short distance upstream.

Illustrations of the serious errors which can result from not conforming to the requirements of the standard can easily be found. In such instances even site calibrations may not resolve the problem since such conditions can produce fluctuations which make accurate measurement impossible. Many experimental studies have been made to try to deal with the corrections required to cater for pulsating flows and these are indicated in the earlier edition of the British Standard BS 1042 (3). The most recent edition of BS 1042 : 1982 is technically equivalent to ISO 5167 and does not include pulsating flows. However further parts of the new BS 1042 are to be published within the next year or so - these will give updated information from Reference 3 not included in Part 1 : 1982 and also deal with the other pressure difference devices referred to earlier.

Because of the square law relation between pressure difference and flowrate, the rangeability of a differential pressure meter is limited normally to 3:1 and to about 5:1 at most. This can only be overcome by converting the meter into one capable of multi-range operation. For example, a bank of orifice meters of different diameter ratios can be built in parallel and the flow switched to the one with the right range. In some situations where brief interruption of the flow is permissible, a greatly extended flow range can be obtained by use of multiple orifice plates (5).

Pressure loss caused by the presence of the flowmeter can pose problems and Fig. 2.5 illustrates that orifice plates and nozzles because of their design dissipate most of the energy which creates the pressure difference. Venturi-meters are low-loss devices.

2.6 CONCLUSION

In choosing a flowmeter, there are many factors to consider and among them the question of accuracy is very important. While it is pointless to pay for a higher accuracy than is necessary, a cheap meter that is not accurate may become expensive to operate. Similarly, unless a meter is calibrated and installed correctly, it will not achieve its potential accuracy.

Thus a typical orifice plate designed and manufactured according to a recognised standard, can be expected to give an uncertainty of say 1-1.5 per cent at maximum flowrate under ideal conditions depending on its diameter ratio. This could well be increased to more than 2.5-3 per cent because of effects of pipe size and inadequate upstream and downstream lengths. If swirling flow exists then errors may rise to 10 or 20 per cent or more.

By calibrating a differential pressure meter an uncertainty of ±0.5 per cent should be obtainable, but this will depend on the quality of the output signal.

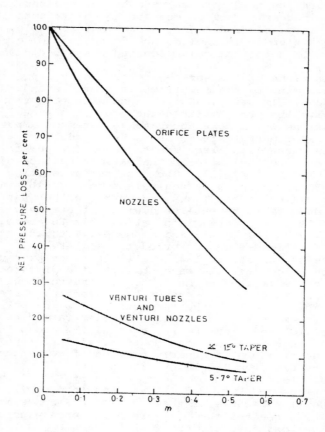

Fig.2.5 Net pressure loss as a percentage of pressure difference

It must also be noted that the standards only apply to flowmeters correctly manufactured and installed and in the same condition as when originally commissioned. As time goes on significant errors can be introduced because of ageing. Thus the sharp edge of the orifice plate can be eroded, the pipe can become too rough, debris can build up against the face of the orifice plate. Excessive flows can cause dishing of the orifice plate while externally the pressure difference transducer may shift its calibration.

Periodic inspection of the orifice plate and internal conditions in the pipe therefore are thus imperative, the frequency depending upon the cleanliness and characteristics of the fluid and materials used.

With all these precautions being observed however, the orifice plate flowmeter can be claimed to be the most reliable and the most predictable flow measuring device at present available.

ACKNOWLEDGEMENTS

This chapter is published by permission of the Director, National Engineering Laboratory, Department of Trade and Industry, East Kilbride from which Laboratory the author recently retired.

© It is Crown copyright, January 1984.

REFERENCES

1. International Organisation for Standardisation, 1980, Measurement of fluid flow by means of orifice plates, nozzles and venturi tubes inserted in circular cross-section conduits running full, ISO 5167 - 1980, Geneva.

2. SPENCER, E. A., 1982, Int. J. Heat and Fluid Flow, 3(2), 59-66.

3. BRITISH STANDARDS INSTITUTION, 1964, Methods for the measurement of fluid flow in pipes, Part 1: Orifice plates, nozzles and venturi tubes, BS 1042 : Part 1 : 1964.

4. International Organisation for Standardisation, 1978, Estimation of the uncertainty of a flowrate measurement, ISO 5168, Geneva.

5. VOSS, L. R., 1974, Flow -its measurement and control in science and industry, Vol. 1, Part 2, 507-515, Instrument Society of America.

Chapter 3

Other flow measuring devices

Dr. M. L. Sanderson

3.1 INTRODUCTION

This chapter deals with flowmeters not covered under the headings of differential pressure devices and electromagnetic flowmeters which have been covered elsewhere. The main emphasis will be on volumetric liquid flow measurement although some reference will be made to gas flow measurement. The literature is replete with novel flowmeters designed to meet specific applications, many of which will never enjoy wide industrial acceptance. One estimate puts the number of commercially available flowmeter types at 100 with the number constantly increasing. Table 3.1 (at the end of the chapter) shows some of the physical phenomena currently employed for flow measurement with their realisations. This chapter concentrates on flowmeters which have been used successfully and widely in industry. The flowmeters to be covered are positive displacement; turbine; transit time, Doppler and correlation ultrasonic techniques; variable area; vortex shedding. Particular emphasis is laid upon the ultrasonic technologies which at the present time represent one of the major growth areas in flow measurement, particularly since they offer the possibility of non-invasive flow measurement with the transducers mounted on the outside of the pipe. For a wider view of flowmeters available the reader is directed to Dowden (1), A.S.M.E. (2) and Hayward (3). Brain (4) provides a survey of mass flow measurement techniques. For a regular update on flowmetering the Fluid Flow Abstracts produced by B.H.R.A. Fluid Engineering provide a useful source of material.

3.2 POSITIVE DISPLACEMENT FLOWMETERS

These flowmeters measure flow quantity as opposed to flowrate in that they deliver a known volume of fluid a measured number of times within the interval for which the flow is to be measured. The known volume can be produced by several means. For liquid flow this can be by means of a semi-rotating piston, reciprocating piston, mutating disc, or gearing arrangement and for gas flow this can be by means of a liquid seal, a diaphragm, or a rotating diaphragm. Some of these are shown in Fig.3.1. In the

sliding vane positive displacement flowmeter shown in
Fig.3.1a the set of vanes rotate within a casing, the
rotation of the vane being caused by the flow of the liquid.
The defined volume is the volume enclosed between two such
vanes, the vanes being arranged in such a way as to provide
sealing with the casing and so limit leakage. The liquid
monitored should be clean since dirt or grit within the
flow can damage the sealing capability of the vanes. The
rotation of the vanes is usually recorded by means of a
mechanical counter.

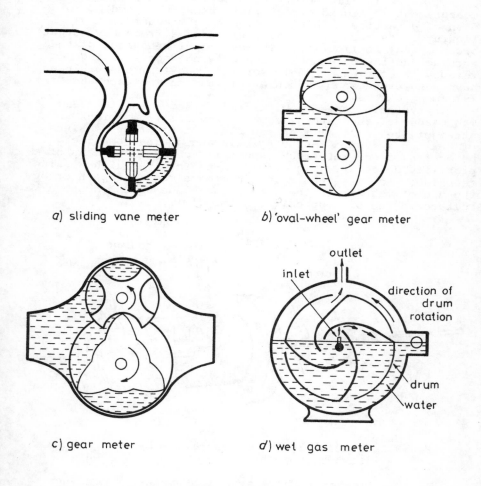

Fig.3.1 Positive displacement meters

Positive displacement flowmeters for liquids are used
extensively in water metering and in the measurement of
petroleum products. They are capable of achieving accuracies of order of ± 0.2% of totalised flow with flow ranges

from $5 \times 10^{-6} m^3/s$ to $0.5 m^3/s$ at pressures up to $10^4 kPa$ and temperatures up to 300°C. They are sensibly insensitive to upstream conditions or to the properties of the fluid being metered. For further details of the operation of positive displacement flowmeters see A.S.M.E. (2), Brain and McDonald (5), Linford (6) and Walker (7).

3.3 TURBINE FLOWMETERS

The turbine flowmeter which can be used for the measurement of liquid or gas flows uses the speed of rotation of a turbine within the flow as a measure of the flowrate. As such, the device produces a frequency output proportional to flowrate. The dynamical behaviour of the turbine and the balance between driving and retarding torques is complex. Models for the behaviour of turbines have been produced by Ruben et al (8) and Thompson and Grey (9) among others.

A typical liquid flow turbine is shown in Fig.3.2. The blades of the turbine come in several forms with straight, helical, or T-shaped blades and various designs of bearing are used including ball and journal bearings. Detection of the rotational speed is achieved by means of a proximity detector. Magnetic pick-up involves either using the blades to alter the reluctance of a magnetic path or by the use of magnets in the tip of the blade which induce a voltage in the pick-up coil (Miller (10)).

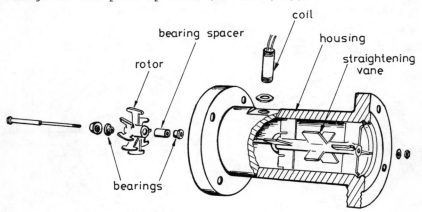

Fig.3.2 Turbine meter (by courtesy of the Foxboro Company)

Turbine flowmeters have a fast response to flow and are capable of providing a rangeability from 10:1 to 20:1 (rangeability is the ratio of the highest to lowest flow measured within a certain error band). Calibration accuracies of ± 0.25% of reading can be achieved with repeatabilities of ± 0.1%. Typically turbines range in size from 6 to 600 mm. They can operate at temperatures of up to 260°C and pressures up to 2×10^4 kPA.

At the low end of the flow range the reading is affected by the viscosity of the fluid, the pick-off technique, and bearing wear. Overspeeding of the turbine can also detrimentally affect its performance. In order to achieve the specified accuracy of the flowmeter it is usually necessary to have a specified length of straight run upstream and downstream of the flowmeter. Typically these are specified as 10 diameters and 5 diameters respectively. For high accuracy applications for use in custody transfer situations it is usual to employ turbine flowmeters with a prover system which provides a regular calibration of the turbines.

3.4 ULTRASONIC FLOWMETERS

A wide variety of liquid flowmeters for use in closed conduits have been designed employing ultrasonics, the most common flowmeters using transit time, Doppler and correlation techniques. Reviews of ultrasonic flowmetering techniques have been provided by McShane (11), Lynnworth (12) who has provided an extensive review of ultrasonic flow measurement with particular emphasis on the ultrasonic aspects as opposed to the electronic aspects and Sanderson and Hemp (13) who have provided a review of the state-of-the-art in liquid flow measurement using transit time and Doppler techniques. Other applications of ultrasonics to the measurement of liquid flow include the measurement of the bending of the ultrasonic beam caused by the flow of liquid (Peterman (14)), and the detection of vortices in a vortex shedding flowmeter. (Joy and Colton (15), Colton (16)) Within open channel flow and river flow measurement transit time techniques have been employed (Genthe and Yamamoto (17), Drenthen et al (18)). Ultrasonics are also widely used as the height of determining element in open channel flow measurement employing flumes or weirs.

3.4.1 Transit Time Flowmeters

These measure the time difference between ultrasonic beams transmitted upstream and downstream in the liquid and as such are designed for use with homogeneous fluids. They have been used for the measurement of both liquids and gases for which they are capable of measuring mass flow rate as well as volumetric flow rate. (Moffat and Fetterhoff (19), Baker and Thompson (20)).

If the liquid in Fig.3.3 is moving with velocity v at angle θ to the ultrasonic beam then

$$T_{12} = \frac{d}{\sin \theta \, (c - v \cos \theta)} \quad \quad \ldots\ldots (3.1)$$

and

$$T_{21} = \frac{d}{\sin \theta \, (c + v \cos \theta)} \quad \quad \ldots\ldots (3.2)$$

where T_{12} is the transit time from transducer 1 to

transducer 2, T_{21} is the transit time from transducer 2 to transducer 1 and d is the diameter of the pipe.

Now since $c^2 >> v^2 \cos^2\theta$, the time difference ΔT is given by

$$\Delta T = T_{12} - T_{21} = \frac{2d.\cot\theta.v}{c^2} \qquad \ldots\ldots (3.3)$$

i.e. the time difference is proportional to v.

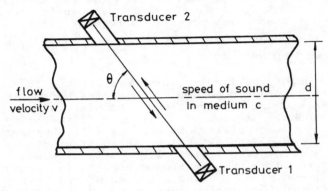

Fig.3.3 Transit time ultrasonic flowmeter

For water flowing in a 100 mm pipe at 1 m/s with the two beams transmitted at an angle of 45° to the flow, the transit time is 94.3μs and the difference in the transit times is only 88 ns. Since ΔT is also proportional to d such measurements are usually restricted to larger pipe sizes and higher velocities. Measurement in smaller pipe sizes is usually achieved by transmission of the beam axially along the pipe or by the use of multiple reflections along the pipe as shown in Fig.3.4.

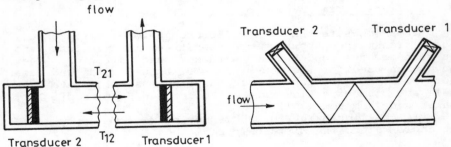

(a) Axial Flowmeter   (b) Multiple Reflections

Fig.3.4 Transit time techniques for measurements in small tube sizes

Transit time flowmeters can be used either with wetted sensors in which the transducers are in contact with

the flowing liquid or as a clamp-on device in which the
transducers are clamped externally to the pipe. Under such
conditions the angle θ and the required separation of the
transducers are then dependent on refractions at the wedge/
wall, wall/liquid, interfaces. The two types of sensor are
shown in Fig.3.5.

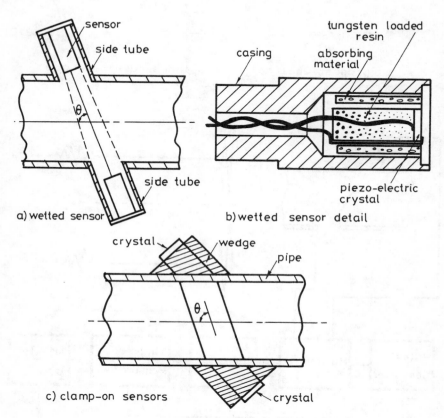

Fig.3.5 Sensors for transit time flowmeters

3.4.1.1. Measurement Techniques.

The two commonly used
measurement techniques for transit time flowmeters are
direct transit time measurements, which are often referred
to as leading edge or pulse techniques, and sing-around
techniques. Fig.3.6 shows a leading edge technique in which
which the two piezo-electric crystals are used both as
transmitters and receivers, the role change being affected
by the multiplexer. A pulse is applied to one transducer,
and the time for its arrival at the other is measured.
The system thus sequentially measures T_{12} and T_{21}. The
defining equation for ΔT shows a dependence on $1/c^2$ and
thus the velocity of sound compensation is essential for
accurate measurement. Water at 20°C shows a velocity of

sound temperature coefficient of +0.2%/°C and thus a wetted sensor device would have a temperature coefficient from this effect of -0.4%/°C.

$$\text{Now } T_{12} \cdot T_{21} = \frac{d^2}{\sin^2\theta(c^2 - v^2\cos^2\theta)} \qquad \ldots\ldots (3.4)$$

$$\text{and thus } \frac{\Delta T}{T_{12}T_{21}} = \frac{2 \cdot \sin\theta \cdot \cos\theta \cdot v}{d} \qquad \ldots\ldots (3.5)$$

i.e. it is proportional to v and independent of c.

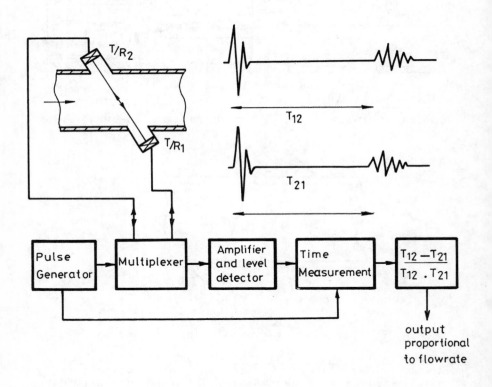

Fig.3.6 Leading edge system

Sing-around techniques provided an output which does not require velocity of sound compensation and they operate as shown in Fig.3.7. The received pulse is used to trigger another pulse at the transmitter and the frequency f' of the resulting pulse train is measured. The role of the transducers are reversed and a new frequency f'' is measured.

$$f' = \frac{\sin\theta(c + v\cos\theta)}{d} \qquad \cdots\cdots (3.6)$$

$$f'' = \frac{\sin\theta(c - v\cos\theta)}{d} \qquad \cdots\cdots (3.7)$$

$$\text{and thus } f' - f'' = \frac{2.\sin\theta.\cos\theta.v}{d} \qquad \cdots\cdots (3.8)$$

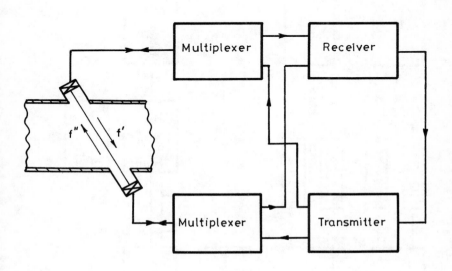

Fig.3.7 Sing-around technique

One of the major difficulties experienced by sing-around techniques is that any obstruction between transmitter and receiver will cause errors in the sing-around frequency. Muston and Loosemore (21) and Hoene (22) have described a technique by which the problem can be eliminated. Fig.3.8 shows the pulse phase comparison technique. The pulses are generated by division of the output of the voltage controlled oscillator. The first is transmitted through the liquid and is then phase compared with the second. Depending on which arrives first the voltage applied to the voltage controlled oscillator is adjusted. In the absence of a received pulse the voltage is maintained at its previous value. In this way it is claimed that only 2% of the pulses need be received in order to achieve the required accuracy. Modifications to the sing-around technique to allow it to work in a clamp-on mode have been made by Yada et al (23). These modifications overcome the problem of a varying angle θ which arises as a consequence of refraction effects.

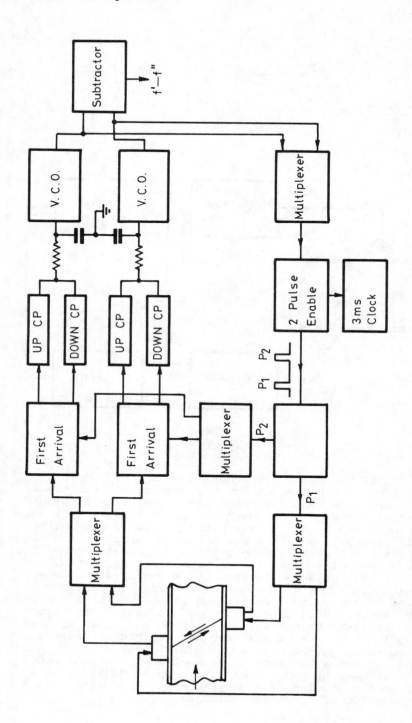

Fig. 3.8 Two pulse phase comparison method

3.4.1.2 Velocity profile sensitivity.

A single beam transit time flowmeter estimates the flowrate by measuring the average flow velocity in the direction of the beam across the length of the beam. Consequently, if it is placed downstream of a bend or valve then, since the fluid velocity across the beam does not represent the average velocity in the pipe, the flowmeter will read in error. Changes in sensitivity of the flowmeter will occur with fully developed flow in long straight pipes as the flow changes from laminar to turbulent. Estimates can be made of the changes in sensitivity due to these effects as follows :-

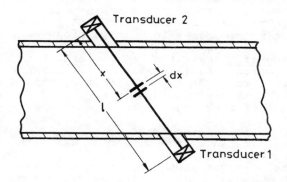

Fig.3.9   Velocity averaging along beam

If v is the component of the fluid velocity along the sound path and in the direction of the sound travelling from transducer 2 to transducer 1, as shown in Fig.3.9, then at any distance x the sound is moving towards transducer 2 at a velocity given by $c + v(x)$, since in general v is a function of x. Thus the total transit time T_{21} is given by

$$T_{21} = \int_0^\ell \frac{dx}{c + v(x)} = \frac{\ell}{c}\left(1 - \frac{\bar{v}}{c}\right) \qquad \ldots\ldots (3.9)$$

where $\bar{v}$ is the average liquid velocity in the direction of the beam along its length ℓ and thus

$$\Delta T = \frac{2\ell\bar{v}}{c^2} \qquad \ldots\ldots (3.10)$$

This mean velocity measurement is not the same as flowrate. As a result an error of approximately 30% can occur in moving from the turbulent regime to the laminar regime (Fronek (24)), and a change in sensitivity of approximately 3.5% occurs in smooth pipes as the Reynolds number is changed from 10^7 to 10^4 (11). The effect of upstream piping such as valves and bends has been estimated by Al-Khazraji et al (25) who indicate that large errors can occur with single beam devices. By using a number of parallel ultrasonic beams and averaging the measured mean velocities associated with each, the effect of upstream fittings can be reduced. In this way an approximation to the mean velocity over the whole cross section can be obtained. Optimization of the position of the beams and the weighting factors applied to the measured mean velocities gives rise to various schemes, analogous to different methods of numerical integration, to make best use of a limited number of beams (Malone et al (26)). Lynnworth and Peterson (27) and Lynnworth (28) describe an alternative scheme to the multi-beam method which involves employing a uniform broad beam of ultrasound extending over the entire cross section. The small diameter tube axial flowmeter is also a transit time flowmeter in which a broad beam is used.

3.4.1.3 Accuracy of transit time flowmeters. Little independent calibration data is available for transit time ultrasonic flowmeters. The typical accuracy quoted by a manufacturer for a sing-around system employing wetted sensors is $\pm$ 1% of flow from 1 to 12 m/s and $\pm$ 0.009 m/s from 0 to 1 m/s in a 75 mm pipe. In larger pipes from 100 to 600 mm the quoted accuracy of $\pm$ 1% of flow extends over a wider range from 0.3 to 12 m/s with an accuracy of $\pm$ 0.00455 m/s from 0 to 0.3 m/s. The repeatability for these larger pipe sizes is quoted as $\pm$ 0.3% of flow for flows above 0.3 m/s. The electronic package of the flowmeters is typically capable of operating in an ambient temperature of between $-30^{\circ}C$ and $55^{\circ}C$ and the temperature limits on the process are from 0° to $84^{\circ}C$ (Sparling Envirotech (29)).

The accuracy limits quoted for clamp-on flowmeters are usually somewhat wider, reflecting the greater uncertainties present in such a device. The typical quoted accuracy for such a device employing a leading edge technique is $\pm$1 to 4% of actual flow for a nominal pipe size, wall thickness, in specified material for velocities above 1 ft/s in non-aerated liquids, with a zero stability of 0.015 ft/s. (Controlotron Corporation (30)). Removal and re-application of the sensors has been found in calibrations made by Barker (31) and Brumer (32) to cause errors of up to $\pm$ 5% in flowmeter sensitivity. Additional sources of error including pipe wall thickness, internal diameter, acoustic velocity in pipe wall material, and transducer axial separation have been identified by Brumer (32). Poor coupling between the transducer and the pipe, and misalignment of the transducer with respect to the pipe axis can cause errors as can the internal surface condition of the pipe.

3.4.2 Doppler Flowmeters

Doppler flowmeters employ scatterers in the flow of provide the necessary frequency shift of the ultrasonic beam and are suited to measurements in liquids in which there are solid particles or gas bubbles to provide the necessary interfaces to scatter the ultrasonic beam. The frequency of the transmitted signal undergoes two Doppler shifting operations and for the configuration shown in Fig.3.10 the frequency of the received signal is given by

$$f_r = f_t \left(1 + \frac{v}{c} \cos \theta_2 - \frac{v}{c} \cos \theta_1\right) \quad \ldots\ldots (3.11)$$

where c is the velocity of sound in the medium
       v is the velocity of the scatterer
and θ_1 and θ_2 are the transmission and reception angles respectively. The Doppler shift, f_d, is thus given by

$$f_d = f_r - f_t = f_t \cdot \frac{v}{c} (\cos \theta_1 - \cos \theta_2) \quad \ldots\ldots (3.12)$$

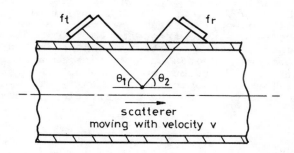

Fig.3.10 Principle of Doppler flowmeter

Fig.3.11 shows an industrial Doppler flowmeter with its associated electronics.

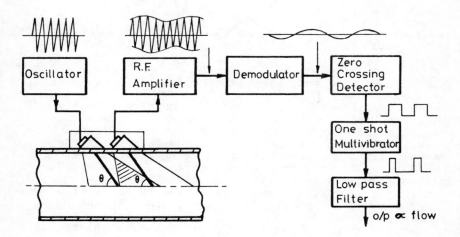

Fig.3.11 Industrial Doppler flowmeter

If all the scatterers are moving with the same velocity v then the Doppler shift, f_d, is given by

$$f_d = 2 f_t \cdot \frac{v}{c} \cdot \cos \theta \qquad \ldots\ldots (3.13)$$

For a flow of 1 m/s in a medium for which the velocity of sound is 1500 m/s, the Doppler shift is 563 Hz if the transmitted frequency is 1 MHz and the beam is transmitted at an angle of 65° to the flow. Application of Snell's Law of refraction gives

$$f_d = 2 f_t \cdot \frac{v}{c_w} \cdot \sin \theta_w \qquad \ldots\ldots (3.14)$$

where θ_w is now the wedge angle and c_w is the velocity of sound in the wedge, i.e. f_d is independent of c.
In general the Doppler shift of the transmitted signal will not be a single frequency but will consist of a band of frequencies with a spectrum as shown in Fig.3.12. The spectrum obtained depends on such factors as velocity profile effects, the distribution of the scatterers, the attenuation of the ultrasonic beam, non axial flow components such as turbulence and the transit time effect of scatterers.

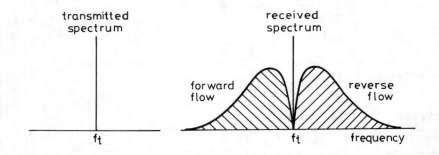

Fig.3.12 Frequency spectra of Doppler shifted signals

The electronics for an industrial Doppler flowmeter are shown in Fig.3.11 in which the Doppler shifted received signal is mixed with the unshifted transmission frequency - this is usually achieved by adventitious leakage of the ultrasonic signal which occurs between the transmitter and the receiver. The received signals are amplified and demodulated. In industrial flow measurement the usual estimate of the Doppler frequency is made by means of a zero crossing detector firing a fixed width monostable every time a zero crossing occurs. It may be shown that such a system gives a measure which is proportional to the r.m.s. frequency of the Doppler spectrum.

3.4.2.1 Limitations and accuracies of Doppler flowmeters

Some of the difficulties and limitations which occur when Doppler systems are applied to industrial flow measurement are likely to occur since: (i) the nature of the scatterers, their size distribution, and their spacial distribution in the flow is likely to be unknown; (ii) the attenuation of the ultrasonic beam and consequently the weighting of the ultrasonic beam is uncertain; (iii) although the Doppler shift is independent of the velocity of sound in the liquid the volume which is being viewed by the flowmeter alters as a consequence of the change in angle of the beam in the liquid; (iv) in large pipe sizes the velocities which are being measured will tend to be very close to the wall and the velocity which will be measured in either turbulent or laminar flow conditions will not correspond to the mean flow and its relationship to the mean flow is unlikely to be known and will vary from one situation to another; (v) the velocity of the scatterers is not the same as the velocity of the fluid. The accuracy of Doppler flowmeters is thus usually rather low. Such devices do, however, have a high repeatability in a given situation. Cousins (33) claims that for a flow range of 0 to 15 m/s a repeatability of $\pm 1\%$ of F.S.D. can be obtained and that for small pipes with well mixed slurries a linearity of $\pm 2\%$ is achievable for Reynolds numbers above 10^5. For large pipe size it is claimed that the accuracy is more dependent on pipe size and Reynolds number. Doppler flowmeters have the advantages that they can be clamped on to the outside of the pipe and that they are cheap. They can operate at temperatures of up to 120°C and at pressures limited only by the pipe work.

3.4.3 Correlation Flow Measurement

Correlation flowmeters estimate the flowrate by measuring the transit time of naturally occurring or deliberately introduced disturbances in the form of eddies or discontinuities between two stations in the pipe shown in Fig.3.13. The method used to estimate the transit time between the two stations is to perform cross-correlation on the signals generated by the disturbances at the two points and this depends for its operation on the persistence of the distribution of the disturbance within the pipe as the liquid flows between the two stations. These disturbances or noise signals can be in the form of thermal or optical noise caused by changes in the temperature of the fluid or changes in its optical density or reflectance, or they can be changes in the electrical impedance measured between two electrodes at each station in the flow. Ultrasonics are commonly used as the method of detecting the disturbances (Coulthard (34), Fell (35)). The ultrasonic transducers have the advantage that they can be placed on the outside of the pipe.

The cross correlation function $R_{xy}(\tau)$ is given by

$$R_{xy}(\tau) = \lim_{T \to \infty} \frac{1}{T} \int_0^T x(t - \tau) \cdot y(t) \cdot dt \quad \ldots \ldots \quad (3.15)$$

The cross correlation function $R_{xy}(\tau)$ will have a maximum value given by $R_{xy}(\tau_{max})$ as shown in Fig.3.13 and an estimate of the flow velocity can be given by $v = \frac{d}{\tau_{max}}$ where d is the separation between the sensors.

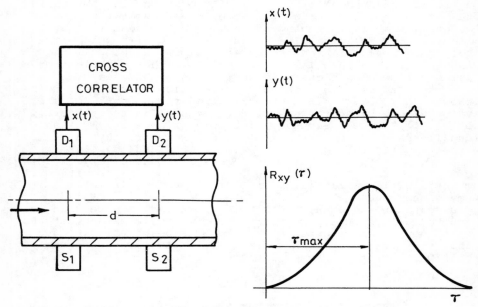

Fig.3.13 Cross correlation flow measurement

The correlation may be performed using analogue or digital techniques. Several schemes are available for performing the cross correlation using microprocessor systems. (Coulthard and Keech (36), Henry (37)). A review of correlation techniques has been given by Beck (38).

3.5 VARIABLE AREA FLOWMETERS

Differential pressure devices such as orifice plates or venturis are constrictions of fixed area across which the pressure drop is measured. Variable area flowmeters employ some method of varying the cross section through which the flow passes in order to maintain constant pressure drop across the cross section. An element within the flowmeter changes its position with the flowmeter to present the flow with a varying cross section. This can be achieved by means of a tapered tube and float meter, a cylinder and piston meter, or an orifice and plug meter. (Miller (10)).

One of the most commonly used forms of variable area flowmeter is shown in Fig.3.14, a tapered tube and float meter. A feedback mechanism ensures that the float is positioned in the tube such that there is a force balance on the float between the differential pressure across it caused by the flow together with the buoyance force acting on the float, and the gravitational force on the float. Any imbalance between the forces will cause the float to move upwards or downwards.

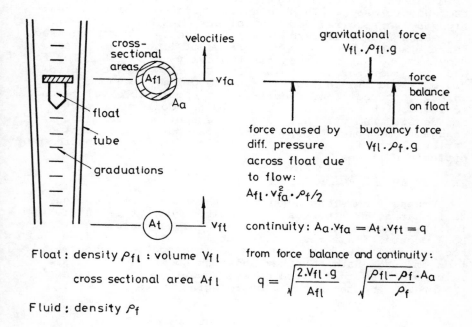

Fig.3.14 Variable area flowmeter

As shown in Fig.3.14 , the balance gives rise to the defining equation of the flowmeter as

$$q = K \sqrt{\frac{2 \cdot V_{fl} \cdot g}{A_{fl}}} \cdot \sqrt{\frac{\rho_{fl} - \rho}{\rho_f}} \cdot A_a \qquad \ldots \ldots (3.16)$$

where
- q is the flowrate
- A_a is the cross sectional area of the annulus at which equilibrium will occur
- A_{fl} is the cross sectional area of the float
- V_{fl} is the volume of the float

$\rho_{fl}\rho_f$ are the densities of the float and fluid respectively
g   is the acceleration due to gravity
K   is a constant introduced to provide correction for the factors neglected in the simple analysis.

If A_a is made proportional to the height of the float in the tube then the flowmeter will have a linear indication. Furthermore, if the tube is constructed from glass then the float level can be read by eye. In cases where, from process considerations, the tapered tube is constructed of metal the position of the float can be detected by electromagnetic means. This technique is also employed when signals are required for control purposes.

Such flowmeters are relatively simple and can be used for the measurement of a wide range of liquids and gases. They can operate at pressures typically up to 3.5×10^6 Pa and at temperatures up to 350°C. Gas flowrates of up to 0.5 m^3/s and liquid flowrates up to 0.1 m^3/s can be measured. Accuracies of order $\pm$ 1% of upper scale value can be achieved with a rangeability of 10:1. Correction factors for density and viscosity should be applied should these change from those for which the flowmeter was calibrated. Variable area flowmeters show little effect from upstream piping conditions.

3.6   VORTEX SHEDDING FLOWMETERS

These flowmeters depend for their operation on the fact that as flow passes over a bluff body it is unable to follow the contours of the body and flow separation from the body occurs. This results in vortices being shed from the body alternately from each side of the body and the generation of a Von Karman vortex sheet as shown in Fig.3.15. This is a phenomena which causes a waving of the flags in the breeze and the singing of telegraph wires.

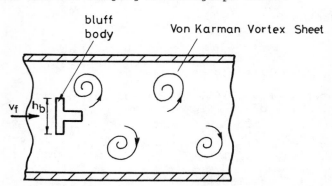

Fig.3.15   Vortex shedding flowmeter

The Strouhal number, S, relates the shedding frequency to the free velocity in that

$$S = \frac{f_{sh} \cdot h_b}{v_f} \qquad \qquad \ldots\ldots (3.17)$$

where f_{sh} is the shedding frequency
h_b is the height of the barrier
v_f is the free velocity.

Linearity of the flowmeter is thus ensured by constancy of the Strouhal number and it has been found that the shedding frequency is directly proportional to the free velocity over a wide region of flow, excluding the laminar region. The range of operation of a particular bluff body and its linearity are dependent upon its shape, triangular or T-shaped bodies being those most commonly used. (Burgess (39)), Miller et al (40)). Detection of the pressure or velocity variations caused by the vortices is by means of ultrasonic, thermal or pressure sensors.

Vortex shedding flowmeters have no moving parts, provide less obstruction to the flow than an orifice plate, and are not particularly susceptible to wear. They can be used for the measurement of liquids and gases and since the Strouhal number is independent of the density of the fluid being monitored they maintain their calibration factor whether being used with liquids or gases. For Reynolds numbers greater than 10^4 they can provide an accuracy of $\pm$ 1% of reading over a range of 20:1. Vortex shedding flowmeters with operating temperatures of up to 200°C and operating pressures of up to 10^4 kPa are available.

Table 3.2, at the end of the chapter, shows a comparison of flow measurement capabilities of the flowmeters discussed in this chapter.

REFERENCES

1. Dowden, R.R., 1972, 'Fluid flow measurement: A Bibliography'. B.H.R.A., Fluid Engineering, Cranfield.

2. A.S.M.E., 1971, 'Fluid meters, their theory and application, A.S.M.E., New York.

3. Hayward, A.T., 1979, 'Flowmeters: A basic guide and source book for users', MacMillan, London.

4. Brain, T.J.S., 1969, 'Mass flow measurement methods', Metron, 1, (1), 1-6.

5. Brain, T.J.S., and Macdonald, L.M., 1977, 'Methods of flow measurement in closed conduit systems' In 'Technical data on fuel' 7th edn., Scottish Academic Press, Edinburgh, 112-126.

6. Linford, A., 1961, 'Flow measurement and meters, 2nd edn., E. & F.N. Spon Ltd., London.

7. Walker, R.K., 1966, 'A historical review and discussion on the design features of positive displacement gas meters', Instrum. Control Syst., 39, (10) 141.

8. Rubin, M., Miller, R.W., and Fox, W.G., 1965, 'Driving torques in a theoretical model of a turbine meter', J. Basic Eng. Ser.D, 87, 2, 413-420.

9. Thompson, R.E., and Grey, J., 1970, 'Turbine flowmeter performance model', J. Basic Eng. Ser.D, 92, 4, 712-723.

10. Miller, R.W., 1983, 'Flow measurement engineering handbook', McGraw-Hill, New York.

11. McShane, J.L., 1974, 'Ultrasonic flowmeters' in Flow, its measurement and control in science and industry, I.S.A., Pittsburg, Pennsylvania, 897-915.

12. Lynnworth, L.C., 1979, 'Ultrasonic flowmeters' Physical Acoustics Volume 14, W.P. Mason and R.N. Thurston, eds., Academic Press, New York, 407-525.

13. Sanderson, M.L. and Hemp, J., 1981, 'Ultrasonic flowmeters - a review of the state of the art', Proceedings of International Conference on Advances in Flow Measurement Techniques, Warwick, England, Sept. 1982, B.H.R.A., Fluid Engineering, Cranfield, 157-178.

14. Peterman, L.A., 1959, U.S. Patent No. 874, 568

15. Joy, R.D. and Colton, R.F., 1972, U.S. Patent No. 3,680,375.

16. Colton, R.F., 1972, 'Vortex anemometry' Proceedings of 20th International Instrument Symposium', I.S.A., Pittsburg, Pennsylvania.

17. Genthe, W.K., and Yamamoto, K., 1974, 'A new ultrasonic flowmeter for flow in large conduits and open channels', in Flow - its measurement and control in science and industry, R.B. Dowdell, ed., I.S.A., Pittsburg, Pennsylvania, 947-955.

18. Drenthen, J.G., Builtjes, P.J.H., and Vermeulen, P.E.J., 1981, 'The accuracy of the total discharge determined by acoustical velocity measurement', in Flow - its measurement and control in science and industry - Volume 2, 1981, W W Durgin, ed., I.S.A., Research Triangle Park, North Carolina, 531.547.

19. Moffat, E.M. and Fetterhoff, K.I., 1974, 'Large volume flow measurement by sonic techniques' in Flow - its measurement and control in science and industry, I.S.A., Pittsburg, Pennsylvania, 933-945.

20. Baker, R.C., and Thompson, E.J., 1975, 'A two beam ultrasonic phase-shift flowmeter', Proc. of Conference on Fluid Flow measurement in the mid 70's, East Kilbride, Glasgow, 8-10 April 1975, N.E.L., Glasgow, Paper H-4.

21. Muston, A.H., and Loosemore, W.R., 1972, U.K. Patent Application 15554/72.

22. Hoene, E., 1978, 'Ultrasonic flowmeter: frequency difference method' in Flow measurement of fluids, H.H. Dijstelbergen and E.A. Spencer, eds., North-Holland Publishing Co., Amsterdam, 147-151.

23. Yada, H., Konno, M., Kobori, T., Kikuchi, A., 1981, 'A clamp-on ultrasonic flowmeter for high temperature fluids in small conduits', in Flow - its measurement and control in science and industry - Volume 2, I.S.A., Research Triangle Park, North Carolina, 549-553.

24. Fronek, V., 1978, 'Ultrasonic measurement of oil flow in a laminar flow - turbulent flow transition region', in Flow measurement of fluids, H.H. Dijstelbergen and E.A. Spencer, eds., North-Holland Publishing Co., Amsterdam, 141-146.

25. Al-Khazraji, Y.A., Al-Rabeh, R.H., Baker, R.C., and Hemp, J., 1978, 'Comparison of the effect of a distorted profile on electro magnetic, ultrasonic and differential pressure flowmeters', in Flow measurement of fluids, H.H. Dijstelbergen and E.A. Spencer, eds., North-Holland Publishing Co., Amsterdam.

26. Malone, J.T. et al, 1971, U.S. Patent No.3,564,912.

27. Lynnworth, L.C., and Pederson, N.E., 1972, 'Ultrasonic mass flowmeter', Proc. of I.E.E.E. Ultrasonic Symposium, Boston, Mass., 87-90.

28. Lynnworth, L.C., 1975, U.S. Patent No.3,906,791.

29. Sparling Envirotech Ltd., 1978, Product Data Street.

30. Controlotron Corporation, 'Series 240 Clampitron Flowmeter: Bulletin 240-1M Installation Manual'.

31. Barker, R., 1980, 'Ultrasonic clamp-on flowmeter experiments', in notes for a short course on ultrasonic flowmeters, Fluid Engineering Unit, Cranfield Institute of Technology.

32. Brumer, R.F., 1977, 'Theoretical and experimental assessment of uncertainties in non-intrusive ultrasonic flow measurement', N.B.S. Special Publication 484, Proc. of the Symposium on Flow in open channels and closed conduits, N.B.S. Gaithersburg, M.D. 23-25, Feb.1977, N.B.S. Washington, 277-291.

33. Cousins, T., 1978, 'The doppler ultrasonic flowmeter' in Flow measurement of fluids, H.H. Dijstelbergen and E.A. Spencer, eds., North-Holland Publishing Co., Amsterdam, 513-518.

34. Coulthard, J., 1975, 'The principle of ultrasonic cross-correlation flowmetering', Measurement and Control, 8, T11-16.

35. Fell, R., 1979, 'A new ultrasonic correlation flowmeter' Proc. of Fluid Mechanics Silver Jubilee Conference, N.E.L., Glasgow, Paper 9.2.

36. Coulthard, J., and Keech, R.P., 1981, Multichannel correlation applied to the measurement of fluid flow', Proc. of International Conference on Advances in Flow Measurement Techniques, Warwick, England, Sept. 1982, B.H.R.A., Fluid Engineering, Cranfield, 253-266.

37. Henry, R.M., 1979, 'An improved algorithm allowing fast on-line polarity correlation by microprocessor or mini-computer', I.E.E. Conference Digest No.1979/32, 3/1-4.

38. Beck, M., 1981, 'Correlation in instruments: cross correlation flowmeters', J. Physics, 14, 7-19.

39. Burgess, T.H., 1977, 'Flow measurement using vortex principles', Proc. of Symposium - The application of flow measuring techniques, Institute of Measurement and Control, London, 321-339.

40. Miller, R.W. et al, 1977, 'A vortex flowmeter - calibration results and application experiences', Proc. of Symposium - The application of flow measuring techniques, Institute of Measurement and Control, London, 341-371.

TABLE 3.1 Physical principles employed in flow measurements

Flowmeter Type	Physical Principle	Realizations
Differential Pressure	Provide a constriction of the flow across which the pressure drop is measured	Orifice plate, venturi, 'Dall' tube, nozzle, elbow flowmeter, annular orifice, target flowmeter.
Electro-magnetic	Employ Faraday's Law of Electromagnetic Induction to generate an e.m.f. in a conducting fluid moving in a magnetic field	a.c. and d.c. pulsed field magnetic flowmeters
N.M.R.	Use Nuclear Magnetic Spin Resonance of the hydrogen atoms to measure flow by means of a tagging technique	Nuclear Magnetic Spin Resonance flowmeters
Positive Displacement	Dispense a known quantity of fluid a measured number of times	Positive displacement flowmeters with semi-rotary piston, reciprocating piston, mutating disc, sliding vane, gearing arrangements, liquid seal, diaphragm, rotating diaphragm.
Turbine	Measure the speed of rotation of a turbine in the flow	Turbine flowmeters with straight, helical, or T-T-shaped blades, with ball or journal bearings, with magnetic or modulated carrier pick-off. Servo-turbines
Thermal	Measure the amount of heat required to maintain constant temperature difference between two points upstream and downstream of a heater	Thermal mass flowmeter Low flowrate thermal flowmeter
Ultrasonic	(a) Measure the difference in transit time between two points in the flow for ultrasound propagated upstream and downstream. (b) Measure the Doppler shift of ultrasound from scatterers within the flow. (c) Use ultrasound as the source of noise in a correlation flowmeter.	Transit time flowmeters employing leading edge, sing around, two pulse phase comparison methods with wetted or clamp-on-sensors. Continuous and range-gated Doppler flowmeters. Correlation flowmeters.
Variable area	Provide a variable area such that the pressure drop across it remains constant.	Variable area flowmeters with tapered tube and float, orifice and plug, slotted cylinder, piston.
Vortex Shedding	Employ a bluff body in order to generate vortices within the flow.	Vortex shedding flowmeters with T triangular, or square bluff bodies with vortex shedding detection by means of capacitive thermal or ultrasonic sensors.

TABLE 3.2 Comparison of flowmeters

Flowmeter Type	Nature of suitable liquids	Upper Temp. Limit	Upper Pressure Limit	Accuracy	Sizes	Additional Notes
Positive displacement	Clean liquids	300°C	10^4 kPa	±0.2% of totalised flow	Suitable for flows in the range 5×10^{-6} m^3/s – 0.5 m^3/s	Little disturbed by upstream or downstream conditions or the properties of the fluid being monitored
Turbine	Clean liquids	260°C	2×10^4 kPa	±0.25% of flow with a rangeability of 10:1 to 20:1. ±0.1% repeatability	6–600 mm	Accuracy affected by viscosity of fluid being measured and upstream flow conditions
Ultrasonic (a) Transit Time	Clean liquids	100°C	limited by pipework	±1% of flow with a rangeability of 10:1 ±1–4% of flow for clamp-on sensors	from 25 mm upwards	Measurement affected by upstream conditions. Provides no additional impedance to the flow
(b) Doppler	Liquids with scatterers	120°C	limited by pipework	poor accuracy varying from situation to situation better repeatability	clamp-on sensors for pipe sizes 25 mm upwards	Suitable for liquids in which there are solids or air bubbles to provide scattering of beam
(c) Correlation	Liquids with detectable eddies or disturbances		limited by pipework	±2–3% of indicated flow		
Variable area	Liquids	350°C	3.5×10^4 kPa	±1% of upper scale value with 10:1 rangeability	suitable for liquid flow rates up to 0.1 m^3/s	measurement little affected by upstream conditions. Density correction required for liquids other than calibration liquid
Vortex shedding	Liquids	200°C	10^4 kPa	±1% of flow over a 20:1 range	40–400 mm	for use at Reynolds numbers $> 10^4$

Chapter 4

Two phase flow measurement

Dr. C. N. Wormald

4.1 Introduction

A continuing problem in the process industry is the measurement of difficult flows, such as quality steam and ammonia vapour where the fluid is partially in a fluid state and partially in a gaseous state. Air in water or oil, coal mixed with oil, pulp stock in water or any combination of liquids, gases and solids of differing substances in varying proportions, presents many flow metering problems.

There are two significant differences between nonhomogeneous and homogeneous flows. First the density is not easily determined; second, one phase or more of the components may not be moving at the same velocity as the main flow, and in some cases may actually be flowing along the bottom of the pipe. This is called 'holdup' or 'slip'. Slip is a complex function of viscosity, and particle size, density differences, surface tension, and superficial velocity of each component. The effects of gravity also alter flow pattens - horizontal, vertical and inclined pipe may cause differing relative velocities for the components.

Most flowmeters used for homogeneous fluids generally measure the velocity of the flow. Where mass flow is required, the user simply introduces the density into the calculation. However in the case of difficult flows, a uniform density cannot be assumed and therefore significant errors in mass flow measurement occur. In an attempt to measure velocity, concentration and mass flow of difficult fluids a number of novel methods have been developed and are discussed in the following sections.

4.2 Historical Background.

Pipeline conveying of solids, where the moving fluid medium carries the solids, has been used for over one hundred years. The Chinese conveyed natural gas through bamboo tubes, and the Romans used lead pipes for water and sewage disposal.

Meaurement of two-phase flow and pipe-line conveying can generally be divided where in the case of solids the conveying medium is either liquid or gas. When a liquid is used as the carrier, it is known as slurry conveying When a gas is used, it is know as pneumatic conveying.

4.2.1 Slurry Pipelines.
The first slurry pipelines found in any numbers were mineral conveying systems in the middle of the 19th century. These systems were in the order of a few hundred feet long. The first pipeline of significant length was built in 1914 to transport coal from barges on the River Thames to the Hammersmith power station, a distance of 1,780 feet. The 8" diameter pipe transported 50 tons per hour of 50% coal by weight at 1.2 metre per second. The line operated successfully for 10 years before it was abanoned due to blockage. Although a number of measurement systems were tried for the mass flow-measurement of the coal, no method was found that could be used on-line and could survive the high wear rate and achieve an acceptable accuracy.

Recent installed pipelines cover a wide range of products, from coal and limestone to fish. Pipeline lengths of many hundreds of miles are not uncommon, which involes many thousands of horse-power provided by the necessary pumps.

4.2.2 Pneumatic Pipelines.
The first mention of using air as a conveying medium was in 1810, when a new method of conveying letters and parcels at speeds of up to 100 mph was proposed. However, it was in 1876 when the first practical conveyor was built to remove dust from grinding and buffing machines. Following this, many grain handling systems were built which involved pneumatic conveying for transporting grain form ships to silo. In the 1920's, following the introduction of powerful positive pressure blowers, European coal mines started to use pneumatic pipelines for storing and backfilling.

One of longest pneumatic conveyors to have been built had a length of 7,600 feet and was used to deliver cement to a USA dam project.

4.2.3 Pneumatic Convening - Major features of two phase flow.
In pneumatic conveying many factors which influence the measuring instrument have to be taken into consideration. Installation, Phase Density and moisture content can give special problems which have to be taken into consideration.

4.2.4 Horizontal/Vertical Conveying.
Depending on the orientation of a pipeline, the distribution of the solid particles across the pipe can be markedly different. In a vertical pipe the flow profile is generally symmetrical across the pipe, where in a horizontal pipe there is often a density gradient across the pipe from top to bottom. This effect is particularly pronounced at lower operating velocities and result in significant measurement errors. Instuments which provide an obstruction to the flow should be avoided in both horizontal and vertical conveyors. In a vertical conveyor, particles tend to slow down and can

eventually fall back. In a horizontal conveyor, saltation can occur. This can be a particular problem, where the velocity is low and the phase density is high, and therefore reults in special measurement problems where the bulk of the solids is in the bottom of the pipe and is being carried along by sand dune effect. In general a vertical pipe will cause fewer problems.

4.3 Measurement of Flow Velocity.

4.3.1 Magnetic Flowmeters.
Magnetic flowmeters are often considered to be the only commercially available flowmeter that is suitable for difficult flows. The principle of the device uses Faraday's well know law of Electromagnetic Induction to measure average flow velocity. The metering section is a non-magnetic tube, usually lined with an insulating material, across which a uniform magnetic field is induced through a portion of the tube. The flowing liquid itself is the conductor moving through the magnetic field. As the disc of liquid passes between the electrodes, an E.M.F. is generated, which is then measured using a high input impedance amplifier. For difficult fluids, its suitablity can be summarised in that it does not obstruct the flow. However, it can only be used with electrically conducting fluids, and is very expensive for the larger line sizes.

4.3.2 Cross-Correlation Flowmeters.
Figure 4.1 illustrates the principle of cross-correlation flow measurement. The cross-correlation function is defined to be:

$$R_{xy}(t^*) = \int x(t-t^*)y(t) \, dt$$

It can be shown that this function has a maximum value when the cross-correlation lag t is equal to the transit time t^* of the tagging signals. Hence the flow velocity V is given by

$$V = L/t^*$$

where L is the spacing of the sensors A and B.

The volume flowrate is the product of the velocity, the cross-sectional area A of the pipe, and a calibration factor k which allows for the measured velocity to be related to the mean velocity in the pipe. The volume flow rate Q is obtained from the equation:

$$Q = k.A.L/t^*$$

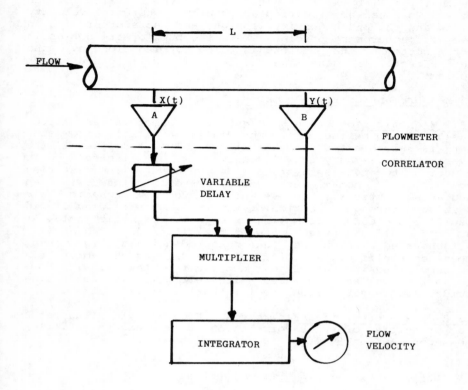

Fig. 4.1 Cross-correlation Flowmeter.

Referring again to Figure 4.1, we can see how simple computing blocks can calulate the cross-correlation function $R_{xy}(t^*)$, determine the transit time z^* of the fluid and calculate the flow rate Q. Although the calculations are simple, a powerful microcomputer is required, because the sensor output data $x(t)$ and $y(t)$ has a fairly wide bandwidth (often 1KHz) and a transit time t^* must be resolved to a high precision (say 99.8%) over a wide range of flows (typically a range of ten to one). In practice, the majority of digital computers and the currently available microprocessors are unable to do the calculations as quickly as required, and this difficulty has had to be overcome in order to make the use of the cross-correlation flow meter a realistic proposition.

The method of flow detection in a cross-correlation flowmeter can cover a wide range of transducers, and many different types have been used to measure flows. These include capacitance, ultrasonic, thermocouples, conductivity, electrostatic and optical transducers. A very important advantage of the type of transducer used in cross-correlation flowmeters is that the transducers are A.C coupled. This greatly simplifies the transducer requirement and avoids any zero-drift problems and the need for re-calibration, which may be due to temperature effects or solids build up on the inner surface of the pipe wall. Transducers are normally mounted in the pipe wall. They do not obstruct the flow or have any parts to wear out. In some measurement situations, where mounting the transducer in the pipe wall is unacceptable, detectors have been mounted on the outside of the pipe. In the case of liquid slurries, a clip-on ultrasonic flowmeter was developed where a single transducer system could fit a wide range of pipe sizes. A typical clip-on flowmeter could be adjusted for pipe sizes 2 to 11 inches diameter.

Flow-velocity measurement using cross-correlation techniques has been under investigation since the mid-60's, Komiya,1966 (1); Kashiwagi and Isobe,1966 (2); Boonstoppel et al, 1968 (3); Bentley and Dawson,1966 (4). Beck,1969 (5), suggested that the natural turbulance eddies be used as tracers. Coultard 1973 (6) implemented ultrasonic transducers for liquid/gas mixtures, and presented a detailed study of the accuracy of the flowmeter. Wormald 1973 (7) developed a 'clip-on' transducer system and measured the flow velocity in large diameter pipes. Ong 1975 (8) extended Wormald's work and measured the velocity of the liquids and slurries using 'clip-on' transducers. All researchers who worked on ultrasonic transducers found it difficult to control the performance of the flowmeter until Battye 1976 (9) used a closed loop control system and demonstrated a method of controlling the flowmeter. He postulated the 'skewness' of the cross-correlation peak and the resultant inaccuracy of the instrument was due to the non-linearity of the demodulator. Using a closed-loop system and maintaining the demodulator at a set crossover point, he claimed successful results. Further work carried out by Leach (10) and Trivedi (11) has confirmed that the main causes of the 'skewness' of the cross correlation peak are the accoustic pathlength variations caused by the instability of the oscillator frequency and the variation in standing wave pattern. A method of controlling the accoutic pathlength between transmitter and receiver has been designed by Balachandran and Beck (12).

4.3.3 The Cross-correlator. A number of cross-correlators suitable for flow-measurement have been developed. In the early days the time delay, which was a measure of the transport time of the flowstream between two axially displaced transducers, was obtanied first from tedious hand calculations, using amplitude measurements from a recorder trace of the two signals. Later, expensive cross-correlators became available, which improved the accuracy and showed that the technique could be used on-line. In order to reduce the cost of the cross-correlator to a competetive cost, when compared with other flowmeters, a number of low-cost correlators were developed. These included an LSI version developed by Jordon (13) and others based on readily available microprocessors Hayes (14), Henry (15), Coultard (16), Fell (17).

4.3.4 Ultrasonic Flowmeters. There are two types of liquid ultrasonic flowmeters. In the first type (time-of-flight meters), a high frequency (approximately 1 MHz) pressure wave is beamed at an acute angle across the pipe. The time taken for the wave to reach the opposite wall depends on whether it is moving with or against the flow and on the speed of sound through the liquid. Flow rate information is obtained from the measured time. In the second type, referred to as the Doppler Flowmeter, the pressure front does not traverse the pipe but is reflected back to a detector by particulate matter moving with the flow. The difference between reflected frequency and fixed transmitted frequency implies the flow rate.

Time-of-flight flowmeters are generally used in clean liquid applications, where the ultrasonic beam is not attenuated or continuously interrupted by fluid or solid particles. Therefore we can conclude that this meter is unsuitable for two phase flows.

The Doppler-type flowmeter, relies on small particles or impurities in the flow, but has also been used sucessfully on many almost-clean as well as dirty streams. Because of the velocity profile, accuracy depends on particle concentration and distribution. Accuracy is also influenced by the relative velocity between particles and fluid.

4.4 Mass Flow Measurement.

In order to appreaciate the problems associated with the measurement of mass flow and difficult flows, it is worthwhile looking at pneumatic conveying of solids and some of the metering methods currently being used.

Transportation of solid material through pipes is a very attractive method of conveying. Coal, foodstuffs, sewage etc are typical of the many materials to be conveyed. In a measurement situation, the flow consists of two component parts, the conveying medium, usually air, and the solid material itself. Apart from having sufficient velocity the conveying medium is usually of little interest, therefore, the function of the flowmeter must be to meter the mass flowrate of the solids component.

4.4.1 Continuous Belt Weighing. Although unsuitable for pneumatic pipes, this system, is one of the most popular method of measuring the mass flowrate of solids materials between two points. The material passing a fixed point on a conveyor belt can be weighed, or the 'loss in the weight' of a hopper can be measured and if necessary integrated with respect to time to give total flow. The belt is suspended by idlers and it is under one set of idlers that a load cell is located. The output from the load cell is proportional to the weight of solids on that part of the belt, and when this is multiplied by the speed of the belt (measured by a tachometer generator in contact with the belt) the mass flow is obtained.

4.4.2 Capacitance Flowmeters-based on flow noise measurement.

Although capacitance cannot be used to measure mass flow directly, it has successfully been used to measure concentration. The principle is based on the changing dielectric property of the solids concentration as the flow passes flush mounted sensor/sensors in the pipe wall. Capacitance is proportional to the dielectric constant (D) of the material between a pair of capacitance plates. However, it is also sensitive to any charges (Q) present:

$$C \propto D.(A/d) \propto Q/V$$

where A - area of plates d - distance between plates

V - voltage between plates

Hence if any charging effects (Q) take place, as in the case encountered in the pneumatic conveying of insulating solids, it will be difficult to reliably say what the capacitance device is sensing. It is also very sensitive to humidity, for instance the dielectric constant of insulating solids is less that ten, whilst water is typically eighty.

Three types of capacitance measurement are commonly used:

1) Static capacitance where the absolute value is measured by a bridge circuit. Commercial bridges are available but require calibration and balancing. The less expensive devices suffer from drift and the operating point has to be frequently adjusted, which makes it unsuitable for industrial applications..

2) Fielden Limited and Tealgate Limited, have manufactured capacitance flowmeters, where the capacitance electrode forms part of a tuned circuit. The principle is that the solid particles in the flow stream pass through the electric field, changing the point of operation of the tuned circuit and hence the frequency of operation of the tuned circuit. A Frequency Modulated (FM) demodulator transforms these frequencies into voltage signals, which can be used for concentration measurement. A full description on this FM device is given by Green (18). See Figure 4.2.

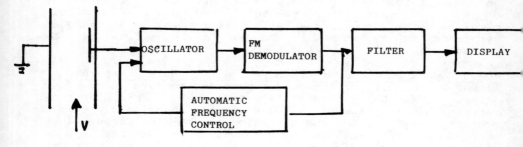

Fig. 4.2   FM Capacitance Transducer for Pneumatic Conveyors.

3) In dense flow applications it is possible that the transport mechanism moves the powder along the bottom or outer wall of the pipe, rather than a uniformly distributed profile across the pipe diameter. To overcome this problem, instead of using one transducer, which will only give local information, Auburn International has introduced a multi-electrode which collects and processes information from six transducers placed around the pipe circumference. The electric field is rotated around the six transducers at 1 KHz and the average dielectric constant is measured and is claimed to be related to the voidage within the flow.

Klinzing (19) has reported this instrument to be a very successful measuring device as well as an instantaneous voidage fraction meter.

4.4.3 Coriolis Meter. This is based on the force exerted by a gas/solid mixture flowing through a U-tube. Micro-motion have commercially developed a mass flowmeter based on this principle. The flow sensor consists of a U-shaped tube which is vibrated at its natural frequency. The angular velocity of the vibrating tube in combination with the 'mass velocity' of the flowing fluid causes the tube to twist. The amount of twist is measured with magnetic position detectors, producing a linear signal which is claimed to be proportional to the mass flowrate of every particle passing through the U-shaped tube. The main advantage of this method is that it is virtually unaffected by variation in fluid or solid properties.

Although this flowmeter, shown in figure 4.3, has many advantages over other devices, it does have a number of disadvantages:

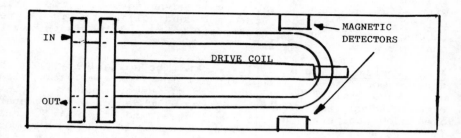

Fig. 4.3. Coriolis Force Principle Flowmeter.

1) The flowmeter is expensive

2) A 2" flowmeter weighs 84 Kgs.

3) A large pressure drop will exist across the device resulting in lower flow rates or increased energy supply costs to maintain previous levels.

4) Servere bends can cause flow separation and slugging behaviour at particular flowrates.

4.4.4 Nucleonic Techniques.

Nucleonic techniques are based on the priciple that the amount of short-level, beta or gamma radiation, absorbed by a material, depends on the mass of material in the path of the radiation. The equation of interest is the Beer-Lambert equation:

$$I/Io = \exp(-\rho \mu \cdot d)$$

where I and Io are respective count rates in the presence of, and in the absence of an obsorber, μ is the extinction coefficient, ρ and d the concentration and thickness of the absorber.

The beta particle attentuation technique, has been used successfully by Howard (20) for the measurement of coal dust concentration under steady running and transient conditions. It requires careful installation and maintenance, but if placed at a position where there is negligible particle impingment on the windows then it can give trouble free operation over acceptable periods.

Problems are encounted with measuring and maintaining μ constant. The technique is subject is subject to errors caused by dust inhomogeneities across the pipe. Therefore the instrument should be placed at a position where the solids concentration is as even as possible. Data processing is tedious and time consuming, therefore commercial apparatus with a data processing facility is advantageous. The basic price of the source, detector and collimator is a few thousand pounds, and the automatic data handling feature will cost extra.

4.4.5 Vortex and Turbine Flowmeters. The suitability and the application of the Vortex and Turbine flowmeter for clean gases and fluids is well understood. However under certain circumstances the Vortex and specially designed flowmeters are used to meter quality (wet) steam and other two phase fluids, such as ammonia, where both liquid and gas (vapour) are present. In these nonhomogeneous applications, mass flow is calculated, by using an effective density. Some manufacturers recommend dividing the gas-phase density by the mixture's quality to yeild a higher density value, under the assumption that the liquid is a mist homogeneously mixed with gas.

4.5 Conclusions

For most single component gases and liquids there are a wide range of commercial flowmeters available. The decision as to which one to purchase often comes dowm to a compromise between cost and the accuracy required. In the case of a flow which is classed as a difficult fluid, it is likely that no commercial flowmeter is available. However, with the introduction of large scale electronics and the development of multi-electrode transducers to deal with complex flows, it is now starting to bring the cost and accuracy to within acceptable limits for industry.

References

1. Komiya K, 1966, Flow velocity measurements by using cross-correlation techniques (1) and (2). Bulletin of the National Research Laboratory of Metrology, Japan. Series 12 64-72

2 Kashiwagi H and Isobe T, 1966. A method for determining the velocity of flow by random tracer injection and correlation techniques (I) and (II). J Soc Inst and Control Eng (Japan),5,(9) 469-481.

3 Boonstoppel F., Veltman B and Vergouwen F, 1968. The Measurement of Flow and Cross-correlation Techniques, Proc Conf on Ind Meas Techniques for On-line Computers, IEE Conf, Publ No 43, 110-124

4 Bentley P.G. and Dawson D. G, 1966, Fluid-flow measurement by Transit-time analysis of Temperature Fluctuations, Trans Soc Instrum Tech, 18, 183-193

5 Beck M S, 1969, Powder and Fluid-flow Measurement using Cross-correlation, PhD Thesis, University of Bradford.

6   Coultard J, 1973, Flow Measurement by Cross-correlation of Ultrasonic Waves, PhD Thesis, University of Bradford.

7   Wormald C N, 1973, Fluid-flow Measurement by Non-contacting Methods, PhD Thesis, University of Bradford.

8   Ong K H, 1975, Hydraulic Flow Measurement Using Ultrasonic Transducers and Correlation Techniques, PhD Thesis, University of Bradford.

9   Battye J S, 1976, An Industrial Cross-Correlation Flow-meter,PhD Thesis, University of Bradford.

10   Leach K G, 1978, A Non-contacting Ultrasonic Cross-correlation Flowmeter for liquids - Accoustic Aspects, PhD Thesis University of Bradford.

11   Trevadi A S, 1978. A Non-contacting Ultrasonic Cross-correlation Flowmeter for Liquids - Electronic Aspects, PhD Thesis, University of Bradford.

12   Balachandran W and Beck M S, 1980, Solid-concentration Measurement and Flow Measurement of Slurries ans Sludges Using Ultrasonic Sensors With Random Data Analysis, Transactions of the Inst of Meas Cont, Vol 2, No 4, 199 - 206

13   Jordon J R, 1973, Correlation Function and Time Delay Measurement, PhD Thesis, University of Bradford.

14   Hayes A M, 1975, Cross-Correlator Design for Flow Measurement; the Design of Digital and Hybrid Cross-Correlators for Time Delay Measurement, and an Analysis of some Sources of Error, with Particular Application to Flow Measurement, PhD Thesis, University of Bradford.

15   Henry R M,1979, On-line Cross-Correlation for Flow Measurement, Journal of Microcomputer Applications, University of Liverpool, Volume 3, No. 3 43 - 51

16   Coultard J, 1984, Six Channel Microcomputer Controlled Correlator, SERC/Inst Measurement and Control Symposium, June 1984, University of Birmingham.

17   Fell R, 1979, Flow Rate, Void Fraction and Bubble Size Measurement in Gas-Liquid Mixtures. PhD Thesis, University of Bradford.

18   Green R. G, 1981, Capacitance Flow Transducers for Multiphase Systems, PhD - University of Bradford.

19   Klinzing G. E, 1981, Gas-Solid Transport, McGraw-Hill.

20 Howard A. V, Development of Techniques for Monitoring the Mass Flowrate of Pneumatically transported Solids, PAC Conference, March 1983.

Chapter 5

Temperature measurement

J. S. Johnston

5.1. THE CONCEPT OF TEMPERATURE AND THE THERMODYNAMIC SCALE

In his opening address to the 1971 Washington Conference on "Temperature its Measurement and Control in Science and Industry" (1), Preston-Thomas told a story of a previous conference during which a delegate was interviewed for the local TV station. "Now tell me Dr 'X'", said the interviewer "in simple words that a layman can understand, just what is temperature?" This was followed by a long and total silence during which the scientist was clearly trying to find an intelligible answer to the question.

Without numerical systems we can rank bodies by degree of 'hotness' (simply by touching them for example) and could construct an arbitrary scale (like Moh's scale of hardness) assigning an unknown temperature a position between two defined standards (hotter than melting ice, but cooler than a human body for example).

We step from this concept to that of 'heat' as a measurable quantity which flows from the hotter to the cooler body. This was the state up to the beginning of the 17th century when the first liquid in glass thermometers were constructed. These, however, had quite arbitrary scales and a further century elapsed before the use of the ice and steam-points as 0° and 100° (or 0° and 80° or 32° and 212°) became common. Some experiments on air thermometers at the start of the 18th century hinted at the existence of a lowest possible temperature.

The concept of heat as a form of motion and temperature as a measure of the intensity of that motion became acceptable in the early nineteenth century and work on steam engines had shown a direct relationship between work done and heat absorbed.

Kelvin eventually put all of this together and defined the absolute scale of temperature in relation to a reversible heat engine working between temperatures T_1 and T_2. The heat provided by the source is Q_1 and that delivered to the sink is Q_2 and these quantities are related by:

$$\frac{T_1}{T_2} = \frac{Q_1}{Q_2} \quad \quad (5.1)$$

(note that T_1 is higher than T_2)
and the efficiency of the engine is clearly:

$$\varepsilon = \frac{Q_1 - Q_2}{Q_1} = \frac{T_1 - T_2}{T_1} \quad \quad (5.2)$$

Numerical values are obtained by defining a single fixed point. This is chosen to be the triple-point of water which is fixed at 273.16K by definition.

In principle then an unknown temperature could be determined by measuring the efficiency of a reversible engine working between the unknown temperature and the triple-point of water.

In practice this measurement cannot be carried out, but gas thermometers with very low gas-densities approximate very closely to the requirement for a thermometer working on the thermodynamic scale as do a few other instruments. None of these techniques is usable as a way of making actual practical temperature measurements or even as a means of routine calibration of practical thermometers. This is not only because the methods are complex and cumbersome, but also because they are not capable of adequate accuracy and precision for industrial and scientific temperature measurement.

5.2. THE INTERNATIONAL PRACTICAL TEMPERATURE SCALE (IPTS)

Because the Thermodynamic scale is not usable directly, a Practical Temperature scale is used based on the temperatures at which well defined changes of state of pure materials occur.

National standards laboratories in various countries determined these temperatures as accurately as possible on the thermodynamic scale and then, for the sake of certainty and uniformity assigned them values which form the fixed points of the IPTS. These are tabulated in (Table 1) below for the most recent version of the IPTS - that of 1968. (2)

TABLE 1. Defining fixed points of the IPTS-68

Equilibrium state	Assigned value of International Practical Temperature	
	T_{68}/K	$t_{68}/°C$
Triple point of equilibrium hydrogen.	13.81	-259.34
Equilibrium between the liquid and vapour phases of equilibrium hydrogen at a pressure of 33 330.6 Pa.	17.042	-256.108
Boiling point of equilibrium hydrogen.	20.28	-252.87
Boiling point of neon.	27.102	-246.048
Triple point of oxygen.	54.361	-218.789
Triple point of argon.	83.798	-189.352
Condensation point of oxygen.	90.188	-182.962
Triple point of water.	273.16	0.01
Boiling point of water.	373.15	100
Freezing point of tin.	505.1181	231.9681
Freezing point of zinc.	692.73	419.58
Freezing point of silver.	1235.08	961.93
Freezing point of gold.	1337.58	1064.43

These temperatures are highly reproducible (to thousandths of a degree in most cases) but they may not co-respond to their temperature on the thermodynamic scale that closely.

For example, the scatter of results used in assigning the temperature of the gold point was a substantial fraction of a degree and on the basis of these results the figure was increased by about 1.4 degree from the 1948 scale to the 1968 scale. Very accurate gas thermometry may now be indicating that the steam point (which has been taken as 100°C for 200 years) may be in error and in some future revision we could have a steam point at (say) 99.97°C.

Between the fixed points particular devices are used as interpolation instruments. Thus between 13.81K (-259.34°C) and 903.905K (630.755°C) the interpolating instrument is the platinum resistance thermometer, from that temperature to 1337.58K (1064.43°C) it is the platinum 10% rhodium/platinum thermocouple and above this last temperature the scale is defined in terms of the Planck radiation law.

In practice any reference to a temperature in whatever units (K, °C, °F, °R etc) is normally assumed to mean temperature as defined by the IPTS. In particular, expressing a temperature in Kelvins does not imply that the temperature is measured on the theoretical thermodynamic scale unless this is specifically indicated.

For most practical measurements for control purposes the distinction is unimportant except that, since the IPTS is revised from time to time to bring it more closely in line with the thermodynamic scale, it may sometimes be necessary to check whether a particular measurement was made before or after a change in the scale. Thus a thermocouple calibrated at the gold-point before and after the change from IPTS 48 to 68 would appear to have drifted by 1.4K, simply because of the change in scale although the emf from the thermocouple at the gold-point might have been identical in the two tests.

5.3. DISSEMINATION OF THE TEMPERATURE SCALE

The IPTS is maintained by certain national standards laboratories (NPL in this country) and by the Bureau International des Poids et Mesures (BIPM) in Paris. These laboratories calibrate suitable instruments supplied to them against the IPTS and supply calibration certificates for them. In particular laboratories of the British Calibration Service (BCS) have thermometers which have been calibrated by NPL and against which they calibrate industrial and other thermometers.

5.4. TYPES OF THERMOMETER

5.4.1. Expansion thermometers.

The oldest form of thermometer and still the most common in everyday use is the mercury or alcohol in glass. While of no direct use in control work they cover a very wide range of temperatures, from say -80°C to 500°C with high accuracy and can, therefore, frequently be used for routine calibration of control devices.

Some expansion devices are used in control applications, mercury in steel, vapour pressure and bimetallic devices being the most common since they are all capable of moving a diaphragm, bellows or shaft. This in turn may operate electrical contacts, a potentiometer or a force-balance system with either an electrical or a pneumatic output.

Although these devices may be losing ground to electrical thermometers they are still fitted in considerable numbers and, because they are simple and self-contained, may often be used in alarms and trips.

The vapour pressure thermometer is particularly useful in dealing with narrow ranges of temperature since the pressure (P) of a vapour in contact with its liquid varies exponentially with temperature (T). i.e. approximately as:

$$P = a\, T^n \exp - \frac{L_o}{RT} \quad \quad \quad \quad \quad (5.3)$$

where a is a constant; n is the difference between the specific heats of the vapour and liquid in units of R the gas constant; Lo is the latent heat of vapourisation.

In practice, a very rough approximation is that the vapour pressure will double for a 10°C rise in temperature.

Note that the vapour pressure in the system will be that corresponding to the temperature of the liquid-vapour interface. If the temperature of the measuring bulb should pass through the ambient temperature the pressure sensitive read-out device will change from being full of gas to being full of liquid; the location of the interface may therefore change, and an error may be introduced by the hydrostatic head of the liquid. Special designs are therefore required for instruments working near ambient temperature.

The scales of vapour pressure thermometers are highly non-linear and there is, therefore, usually a distinction between the total temperature range and the usable temperature range of a particular instrument.

5.4.2. Thermocouple thermometers.

This is not the place to go deeply into the theory of the thermoelectric effect which is in any case imperfectly understood. It may be sufficient to consider that if two metals are connected together at one end, as shown in (Fig.5.1) a potential difference may be measured at the open ends when a temperature difference exists between A and B. This results from the difference in work functions at the hot and cold junctions, and from the potential gradient which must accompany a temperature gradient in a conductor.

Figure 5.1. - Thermocouple

Almost any pair of dissimilar metals could be used to produce a thermocouple, but a limited number have achieved a degree of international acceptance. It should very rarely be necessary to use couples outside table 2 in normal industrial work. (3)

TABLE 2. Types of thermocouples

Type	Composition	Normal Operating Range	Maximum 'Spot' Reading
E	Nickel-chromium/ Copper nickel or Chromel/constantan	-200°C to 850°C	1100°C
J	Iron/copper-nickel or Iron/constantan	-200°C to 850°C	1100°C
K	Nickel-chromium/ Nickel-aluminium or Chromel/alumel	-200°C to 1100°C	1300°C
T	Copper/copper nickel or Copper/constantan	-250°C to 400°C	500°C
B	Platinum 30% Rhodium/ Platinum 6% Rhodium	0°C to 1500 °C	1700°C
R	Platinum 13% Rhodium/ Platinum	0°C to 1400°C	1650°C
S	Platinum 10% Rhodium/ Platinum	0°C to 1400°C	1650°C

These thermocouples have recently been studied at various standards laboratories and revised emf/temperature tables produced. (4).

The difference between the older BSI tables and the new are usually small but may be significant in accurate work. The difference can be as much as 4°C in the case of copper-constantan (Type T). There are also significant differences between these new tables and the older NBS tables for precious metal thermocouples.

The tables also include polynomial representations of the emf-temperature relationship for use in computer data-processing. These polynomials need to be treated with caution; they contain up to 14 terms expressed to 11 significant figures and cannot be truncated to obtain a lower precision representation.

Dr Coates (5) of NPL has published some alternative forms based on Chebyshev polynomials which are much easier to handle.

The important feature of the thermocouple thermometer is that it always measures the difference between one temperature and another. The position of the reference junction (usually called the cold junction) is not always obvious but it must be defined for accurate work.

Temperature measurement   79

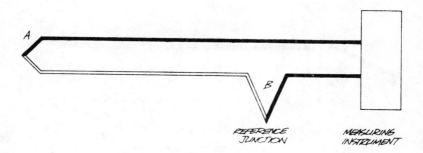

Figure 5.2.   -   Thermocouple with Reference
                    Junction.

In figure 5.2. the position of reference junction is clear and its temperature may be controlled by immersion in an ice-bath or some other region of known temperature. Note however that if the measuring instrument has copper wiring, there are two more junctions at that instrument's terminals - there must be no temperature difference. between these terminals if accurate measurement is required.

In figure 5.3. we have a less clear situation:

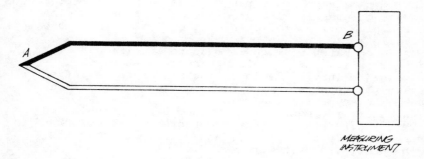

Figure 5.3.   -   Directly Connected Thermocouple

There is no clear reference junction - in fact it is at the instrument terminals and the instrument will, unless it has internal compensation, measure the difference in temperature between the hot junction and its own terminals.

In figure 5.4. a cold junction is provided encompassing both legs of the thermocouple and permitting the use of copper wire back from this point to the instrument. This arrangement is particularly appropriate in scanning and logging applications, since a large number of thermocouples may all have their reference junctions in the same temperature controlled zone with the wiring to the scanner being entirely of copper

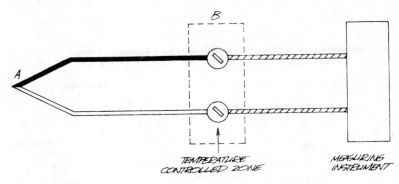

Figure 5.4. - Thermocouple with Local Reference Junction.

In practice, deflection instruments may have bi-metal strips to move the zero of the pointer to compensate for ambient changes while electronic instruments may use resistance thermometers in bridge circuits to provide a signal for cold-junction compensation.

In industrial practice one can seldom use the same wires all the way from the measuring point to the indicator and extension leads or compensating leads may be required. The former are of nominally the same material as the thermocouple, while the latter are alloys having a similar emf-temperature relationship to that of the thermocouple over a limited temperature range. The main application of compensating leads is with precious metal thermocouples where the high price of platinum alloys makes their use essential.

As each batch of wire is made to specified tolerances on its emf-temperature relationship, the various joints inevitable in a long run of cable introduce errors in the total emf measured and these must all be evaluated in any error analysis.

An excellent manual on thermocouples is published by ASTM (6) note however that the thermocouple colour codes shown are U.S. standards and different codes are used in the U.K. and in Europe. (7).

The life of the thermocouple depends on the type, the temperature of use, the immediate environment of the thermocouple wires and their diameter. Long life and very good stability are features of the precious metal couples provided that contamination from iron vapour is avoided at temperatures above 500°C. For base metals used close to their upper temperature limit, a life of a few thousand hours can be expected.

Base metal couples have a typical tolerance on the emf-temperature relationship of ± 0.75% of Celsius temperature or ± 2.5°C, whichever is the greater, and no significant improvement on this is likely to be possible in industrial conditions.

The thermal history of the thermocouple and temperature gradients in the wiring can influence the emf. If accurate calibration is required it is often best done in situ, using a different form of measuring instrument alongside it.

Cycling a base metal couple above about 800°C can cause calibration shifts of a few degrees. The sensitivity to temperature gradients may be illustrated by a calibration on a type K thermocouple performed at NPL at 400°C. This measurement was done first in a furnace and then in a liquid bath, and the results differed by about 6°C.

A new and promising thermocouple has been developed in Australia called "Nisil-Nicrosil" which is similar to a nickel-nichrome couple, but the two limbs contain a small proportion of silicon which oxidises on the surface to give a self-passivating layer, largely eliminating drift through oxidation and hence inhomogeneity.

Thermocouples in swaged mineral insulated cables are generally less affected by inhomogeneity problems than units fabricated from wire and ceramic insulators.

5.4.3. Resistance thermometers (10)

These divide into two main types: thermistors, which are usually sintered mixtures of metal oxides with the characteristics of a semiconductor, and units based on the change in resitivity with temperature of pure metals or alloys (see Fig.5.5).

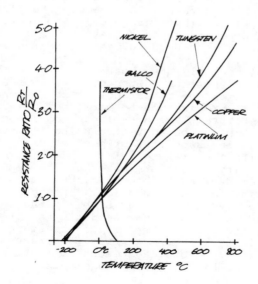

Figure 5.5.   -   Resistance-Temperature Curves for various materials.

Thermistors are not yet in common use in industrial measurement and control, although they are widely used in laboratory work. The reasons for this difference may be found in their highly non-linear characteristics and the lack of standardisation of their resistance-temperature relationships. They are, however, capable of good stability and particular manufacturers will offer units interchangeable with others of their own manufacture within a fraction of a degree.

A particularly good book on thermistors is entitled "Semi-conducting temperature sensors and their applications". (8).

Temperature detecting elements based on the change in electrical resistivity of pure metal wires have a long history, having been proposed by Siemens in 1871 and having been used by Callender for work of the very highest accuracy in 1886. It is only comparatively recently, particularly in this country, that they have been used in really large numbers in industrial practice.

Copper is, of course, an obvious candidate as the sensing material, since it is readily available in fine wires of high purity. Its use, though, is often confined to special purpose or laboratory instruments because of its susceptibility to corrosion and oxidation.

The low resistivity of copper also means that very long lengths of fine wire are needed to produce a useful resistance; the result is that a copper resistance thermometer is normally rather bulky. But for some purposes it has one considerable advantage in that its resistance temperature relationship is the most nearly linear of common pure metals. Its low cost has led to its use in some non-critical but large-scale applications such as car radiator temperature sensing systems. It is also used in long thermometers used to measure average temperature in oil tanks.

Nickel is frequently used, particularly in the United States, in temperature detectors. It has a high temperature coefficient of resistance at room temperature and the slope of its resistance-temperature curve increases with rising temperature. The non-linear resistance temperature relationship can be corrected using additional passive components in associated bridge networks, to give outputs linear with temperature.

Its resistivity is higher than that of copper, permitting smaller devices to be built, but there are still problems with oxidation and corrosion. The upper temperature limit is set by the peculiar shape of the resistance-temperature curve in the neighbourhood of the Curie point at 358°C, and by the instability of the resistance of the element when it is cycled through this temperature. There is little or no international agreement on the resistance-temperature relationship for nickel and this has hindered its wider use.

Platinum is not subject to oxidation or corrosion in the majority of environments; it can be obtained in fine wire of extreme purity and its resistivity is higher again than that of nickel.

It is also capable of covering a wide temperature range; the British Standard (9) shows a resistance-temperature tabulation from -200°C to +850°C.

The industrial resistance thermometer element consists of a small glass or ceramic detector containing the platinum coil inserted into a metal sheath. More recent detectors have a film of platinum deposited onto a ceramic support.

There is now virtually world-wide agreement on the resistance-temperature relationship of industrial units- 100Ω at 0°C and 138.5Ω at 100°C. (Ref.9, Table 1.)

BS 1904 quotes two tolerance grades: Grade A - roughly ±0.2% of temperature in °C and Grade B - roughly ±0.5% of temperature in °C. The usable temperature range is from -200°C to 850°C but special care is needed above 500°C, mainly because of the danger of contamination of the platinum by iron vapour or by metallic elements reduced from the glass or cermaics. Stability is very good - a few hundredths of a degree change after several years at 600°C. Measurements are made with some form of bridge circuit which may form part of a temperature transmitter. The bridge must be designed to reduce the effects of lead resistance. Normal industrial practice uses three-wire connections but a four-wire current and potential lead connection is needed for the highest accuracy. (Ref 10, Figs. 12-17.)

The non-linearity of the resistance temperature relationship is small, and can be ignored for most purposes over spans of a few tens of degrees.

A rough rule of thumb is that the terminal non-linearity is about 0.4% per 100°C span. Thus a thermometer covering 0-100°C would have a non-linearity of 0.4% of 100°C (or 0.4°C); used over the span 0-300°C it would have a non-linearity of 1.2% of 300°C (or 3.6°C.)

5.4.4. Radiation thermometry (pyrometry) (11).

The temperature of a body can be determined by measuring the thermal radiation it emits. In the range of temperatures covered by most applications of measurement and control the wavelengths to be considered run from the far infra-red down to the visible spectrum although work on plasma temperatures involves measurements in the ultra-violet.

The spectral radiance $N_{\lambda b}$ of a black-body is given by the Planck radiation equation:

$$N_{\lambda b} = \frac{C_1}{\pi \Omega_o \lambda^5 [\exp(C_2/\lambda T) - 1]} \quad \ldots\ldots\ldots(5.4.)$$

where C_1 and C_2 are constants, λ is the wavelength and Ω_o the unit solid angle.

This relationship is shown in figure 5.6. and it is this equation which, with a specified value for C_2 is used to define the IPTS above the gold point. The other constant does not require definition since the temperature scale is defined by the ratio of the spectral radiances (both measured at the same wavelength) of a black body at a temperature T and at the melting point of gold.

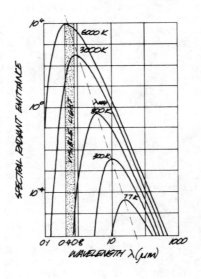

Figure 5.6.   -   Black-body Radiation.

All types of pyrometer use the curves of figure 5.6. although in different ways. Thus a Total Radiation Pyrometer attempts to measure the total area under the curve relating to the temperature of the source. This area is given by the well known Stefan - Boltzman Law:

$$N_b = \frac{\sigma}{\Omega_o \pi} T^4 \quad \dots\dots\dots\dots\dots\dots\dots\dots\dots\dots (5.5)$$

where N_b is the total radiance and σ is a constant.

In practice such instruments tend to be somewhat slow in operation and to be limited in actual bandwidth by the lens or window material and by the detector. The latter is usually a thin film thermopile or bolometer and, because they use all of the available radiated power, these instruments can be used to measure comparatively low temperatures (down to -50°C in some cases).

Narrow band optical pyrometers use filters to limit the band of wavelengths passed to the detector and hence respond according to the Planck equation [5.4]. Some types of laboratory rather than process control instruments, compare the radiation from a standard filament inside the unit with that from the target; equality of the two is achieved by varying the filament current. This current is then used as a measure of the temperature.

Two colour pyrometers compare the spectral radiances at two wavelengths to identify the temperature. These devices are complex and not in common use in process control; they cannot be used if the emissivity of the target is a rapid function of wavelength.

The narrow band pyrometer has a particular advantage in dealing with transparent materials (plastic, glass etc) since it is possible to choose a wavelength at which the material has high absorption and is thus opaque.

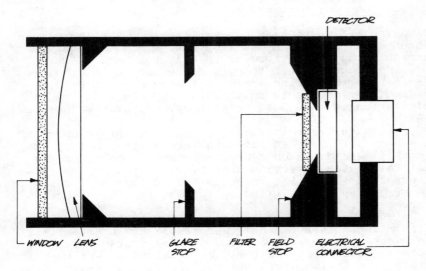

FIG 57

Figure 5.7. shows a typical industrial pyrometer. In practice this may be supplemented by electronics to amplify and linearise the signal.

Calibration of pyrometers is carried out using black-body furnaces at known temperatures or by means of tungsten-ribbon lamps with known relationships between the ribbon temperature and the heating current. The major problem is the wide range of emissivities of the bodies whose temperatures are to be measured. On a single material this may range from say 0.05 for polished aluminium to 0.4 for sand-blasted aluminium, and from 0.07 for polished iron to 0.79 for rolled iron.

Since the emissivity can be represented as a multiplier in front of the expressions in eqs. 5.4. or 5.5. it must be compensated by a change in gain at the measuring instrument or by forming the equivalent of a black-body around the part of the surface in question.

Another problem lies in the possibility of some of the radiated energy being absorbed by fume, water vapour, carbon dioxide etc., in the atmosphere between the target and the pyrometer. This may be minimised by appropriate choice of wavelength, by moving the pyrometer closer to the target (and hence possibly having to water-cool the pyrometer) or by using fibre-optic light-guides to achieve a similar result.

5.5. INSTALLATION AND USE OF IMMERSION THERMOMETERS

5.5.1. Conduction errors (cold-end effect)

The effect of heat conduction along the thermometer element will result in the temperature of the sensing portion being biased towards that at the head.

The magnitude of the effect depends on the heat-transfer to the medium and on the length and thermal conductivity of the thermometer stem. Usually a rough calculation is enough to establish the likely magnitude. (10).

A useful test is suggested in Ref. (9) in which the stem passes through an ice-bath into a steam-bath. The length immersed, which causes no perceptible error, is the "calibrated immersion depth", while that which causes an error of 1°C is the "minimum usable immersion depth". The error may be minimised by lagging the top of the thermometer.

When, as is usual in process control, the thermometer is mounted in a pocket (thermowell) the problem is made worse by the lack of good thermal contact with the pocket. Some improvement may be obtained by spring-loading the thermometer into contact with at least the tip of the pocket.

5.5.2. Self-heating.

This applies to all types of resistance thermometer (including thermistors) and is an error caused by the heating effect of the measuring current. A typical figure for an industrial platinum resistance thermometer element is 0.03°C/mW in water or ice. In laboratory work, with platinum resistance thermometers, 1mA is a typical measuring current while up to 5mA would be normal for accurate industrial work. Check carefully before using any instrument using a measuring current in excess of 10mA. For thermistors the measuring currents may have to be restricted to the order of tens of microamperes. Much larger self-heating figures will occur in slowly moving gases.

5.5.3. Time response.

It is usual to assume that a thermometer element behaves as a single lag system and to declare the time to achieve 63% of a step change as the time constant; usually measured by plunging the sensor into hot water moving at 1 m/s.

Note that the figure obtained is valid only in water at 1 m/s. In air at the same velocity, the time constant might be a hundred times larger. In water at 10 m/s it might be one third of the declared value. (10).

Again the use of a pocket or thermowell may greatly increase the time constant. It is not unusual to find that a thermometer with a time-constant of a few seconds when plunged into water has this figure increased to a few minutes when tested in a pocket.

5.5.4. Thermo-electric Potentials.

A resistance thermometer is inevitably exposed to a temperature gradient along its length which means that any inhomogeneity in the connecting wires will produce a stray thermo-electric potential across the terminals.

BS 1904 requires that the error caused by this spurious e.m.f. shall be small compared with the inter-changeability tolerance when the resistance is measured at about 1mA.

It is, therefore, not advisable in d.c. resistance measurement to use currents below about 1mA.

The equivalent problem in the extension or compensating wires of a thermocouple has already been mentioned.

5.5.5. Total or stagnation temperature (10)

In a high speed gas flow the temperature sensed (T_T) is higher than the normal static temperature (T_S). If the gas were brought adiabatically to rest at the sensor the total temperature would be given by:

$$T_T = T_S [1 + 0.5 (\gamma - 1) M^2] \quad \dots\dots\dots\dots\dots\dots (5.6)$$

where γ is the ratio of specific heats for the gas and M is the Mach No. For air at room temperature (300K) moving at Mach 0.2, this increase is 2.4°C. An ordinary thermometer will usually give a reading part way between static and total temperatures. If this error is serious, special probe designs are required.

It is naturally particularly important to make this distinction in aircraft instrumentation; at Concorde's maximum speed of Mach 2 the total temperature is 170°C above the static temperature, which at the cruising altitude might be -60°C.

5.5.6. Installation and vibration.

A thermometer element may be directly immersed in the medium, if shutting down the plant to remove a sensor for repair or replacement is acceptable, and if fast response is required. It may be subject to accidental damage and will be very prone to vibration.

More usually the thermometer will be mounted in a pocket or thermowell. This must be stressed to stand the line pressure and the sideways thrust of the moving fluid. One must always check for likely vibration and the effect of the mechanical 'Q' of the device causing severe vibration at the tip. Thermometer elements must be supported or spring loaded in the pockets to prevent rattling and consequent early failure.

The commonest cause of vibration is vortex shedding from the sides of the probe. The vortex shedding frequency given by:

$$f = 0.2 \frac{V}{d} \quad \dots\dots\dots\dots\dots\dots\dots\dots (5.7)$$

(where V is fluid velocity and d probe diameter) should be calculated and compared with the fundamental lateral resonant frequency of the probe. A potentially dangerous condition is present if the vortex frequency at maximum flow-rate is greater than the resonant frequency.

REFERENCES

1. "Temperature, its measurement and control in Science and Industry". Vol.4, 1972, Instrument Society of America.

2. National Physical Laboratory: "The International Practical Temperature Scale of 1968", H.M.S.O.

3. BS 1041: Part 4: 1966 "Code for Temperature Measurement: Thermocouples".

4. BS 4937: Parts 1-7: 1973: "International Thermocouple Reference Tables".

5. Coates P.B.: "Functional Approximation to the Standard Thermocouple Reference Tables", National Physical Laboratory, Report QU 46, March '78.

6. American Society for Testing and Materials: "Manual on the use of Thermocouples in Temperature Measurement". Publication STP 470A.

7. BS 1843: 1952: "Colour Code for Twin Compensating Cables for Thermocouples".

8. Sachse H.B.: "Semi-conducting Temperature Sensors and their Applications", John Wiley & Sons Inc., 1975.

9. BS 1904: 1984: "Specification for Industrial Platinum Resistance Thermometer Sensors".

10. Rosemount Incorporated Bulletin No. 9612: A general Technical Bulletin on Platinum Resistance Temperature Sensors.

11. Heimann W., and Mester U.: "Non-contact Determination of Temperature by Measuring the Infrared Radiation emitted from the Surface of a Target", Inst. Phys. Conf., Ser.No.26, 1975.p.219.

GENERAL REFERENCES

12. Baker H.D., Ryder E.A., Baker N.H.:"Temperature Measurement in Engineering", Vols. 1 & 2, John Wiley & Sons.

13. National Bureau of Standards: "Precision Measurement and Calibration - Temperature", United States Dept. of Commerce, Special Publication 300, Vol.2, 1968.

14. Liptak B.G.: "Instrument Engineers' Handbook", Vol.1, 1969, Chilton Book Company.

15. Barber C.R.: "The Calibration of Thermometers", National Physical Laboratory, H.M.S.O.

16. BS 1041: 1943: "Temperature Measurement".

Chapter 6
Pressure measurement
U. Erdem

6.1 INTRODUCTION

The measurement of pressure is probably one of the most important and commonly employed measurement in industry. This is perhaps due to the fact that in many industrial applications flow rate and fluid velocity can be derived from pressure.

Pressure cannot be measured directly but can be deduced by measuring the force acting vertically upon a known area. This is the basic principle of all pressure measurement techniques. It is essentially sensed by a mechanical sensing assembly, such as a diaphragm in a pressure transducer, which presents an accurately defined surface area to act upon.

The unit of pressure is pascal, Pa, and has the units of force per unit area. The engineering quantity stress has the same unit.

All pressure measuring devices respond to a change of differential pressure across them. There are basically four types of measurement configurations.
a. Gauge pressure, called psig, is when the measured pressure is referenced to the ambient atmospheric pressure. The reading is zero when the input pressure port is vented to atmosphere.
b. Absolute pressure, called psia, is when the measured pressure is referenced to full vacuum, usually a sealed chamber within the device. The reading is approximately 101.3 kPa (14.7 psia) when the input pressure port is vented to atmosphere.
c. Differential pressure, called psid, is when one measured pressure is referenced to another pressure.
d. Sealed pressure, called psis, is when the measured pressure is referenced to a pressure, usually in a sealed chamber within the device. This sealed pressure may be the atmospheric pressure in applications where it may not br possible to vent the gauge pressure measuring device due to unsuitable environment.

Although the unit of pressure is pascal, Pa, within the SI system, the most popularly used pressure unit in industry in the UK and USA, still remains primarily lb/in^2 or psi, and perhaps, followed by bar. It is still an accepted engineering practice to quote the pressure of the compressed air used on the shop floor in terms of 80-100 psi not 550-700 kPa or the pressure of the bottled air as 2000 psi not 13.8 MPa.

The pressure measuring instruments can be divided into two main groups. The first group are the direct pressure measuring instruments which determine the value of applied pressure by directly calculating the force applied upon an accurately known area. Various types of manometers and dead weight testers (pressure balances) are in this group. The second group are the indirect pressure measuring instruments that are based on the use of elastic mechanical elements to which the pressure is applied. Some instruments such as Bourdon tubes and capsules are allowed to have comparatively large deflexions in order to drive dial gauges. In pressure transducers, the applied pressure is opposed by a light but stiff diaphragm whose deflexion, usually very small, is sensed by a secondary transducer such as strain gauges, LVDT, capacitive transducers and an electrical output is produced. All the devices in this group are calibrated with one of the instruments in the first group. For the pressures above atmospheric, 100 kPa, a dead weight tester is used, for pressures below this a calibrated mercury column is used.

The National Physical Laboratories has the responsibility for the maintenance and dissemination of the national standards for the pressure measurement in the UK.

As the majority of modern process plants increasingly employ pressure transducers with electrical output for their data processing, only a limited reference will be made to the mechanical pressure transducers.

It will be noticed that since force is the primary input for weight and pressure measurement, the methods of measurement have common techniques. There is, for example, very little difference in the operating principles and manufacturing technique of a low range strain gauge load cell and a pressure transducer using the same force bearing element.

6.2 MANOMETER

This is a pressure gauge using a liquid column as the means of pressure measurement. The measuring principle is based on the hydrostatic pressure relationship that a differential pressure is related to the column differential Δh by the expression, (figure 6.1.A),

$$\Delta p = \Delta h \rho g$$

where ρ is the density of the liquid and g is the gravitational acceleration. This is a linear relationship assuming ρ and g are constants and the accuracy of measurement is limited by the accuracy with which the differential column height Δh can be measured. This may be done visually by the use of vernier graduation or by ultrasonic means or a more sophisticated laser interferometry technique to achieve maximum accuracy. The latter technique is used by the N.P.L. on their long-range primary barometer.

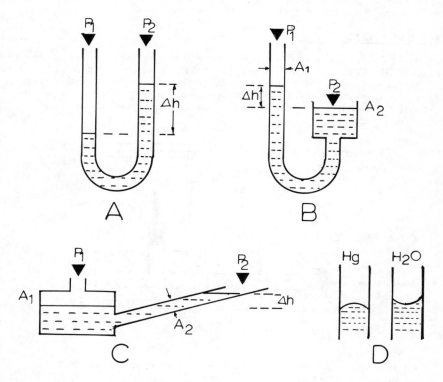

Fig.6.1 Types of manometers: A. U-tube; B. enlarged limb; C. inclined tube; D. meniscus of mercury and water in a tube.

There are basically three types of manometers, U-tube (fig. 6.1.A), enlarged limb (fig. 6.1.B) and inclined tube enlarged limb (fig. 6.1.C).

The liquids used depends on the pressure to be measured. Mercury is the most commonly used liquid due to relatively low temperature expansion characteristics and low evaporation rate. For a Δh = 500 mm difference in columns the pressure required is 66.7 kPa for mercury, 3.9 kPa for alcohol and 4.9 kPa for water.

Although a manometer is a simple apparatus to construct and operate, it is difficult to achieve the full accuracy that it is capable of providing. The readings of column heights, Δh, may need extensive corrective calculations to overcome the temperature effects on the liquid, the tube and the supporting structure. Any contamination of the liquid which may affect its properties and influence of the meniscus (fig. 6.1.D) on the readings should also be taken into account.

The measurement uncertainty of liquid manometers is

generally 0.25%, however, when operated in a controlled environment with sophisticated length measuring techniques, uncertainty levels of $\pm$ 0.01% can be achieved.

6.3 DEAD WEIGHT TESTER

These pressure measuring instruments are normally used in laboratories and standard rooms in industry for calibration purposes. The working principle of the dead weight tester is based on balancing the force exerted by the working fluid on a piston of known area by the set of calibrated weights.

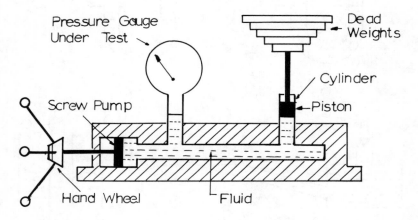

Fig. 6.2. Cross sectional view of a Dead Weight tester.

Fig. 6.2 illustrates the general arrangement for a typical dead weight tester. To operate the instrument, the required dead weights are placed on the weight support table and the handwheel is screwed till the piston carrying the weights floats freely. The piston is then rotated to ensure that the fluid film in the piston-cylinder clearance is uniform and the friction is minimum. The pressure applied to the pressure gauge is then equal to the applied dead weight divided by the area.

The fluid used is normally a mineral oil type selected to suit the piston-cylinder clearance. It is important that this oil should be free from any contamination to avoid scoring the piston and cylinder walls. In order to realise the maximum accuracy from a dead weight tester it is necessary to make a number of corrections. These are for the piston and cylinder deformation as a function of pressure, ambient temperature and bouyancy effects on the dead weights. It is also necessary to take into account the effect of gravitation on the fluid, piston assembly and the dead weights according to the local g value.

The dead weight testers can achieve uncertainty of measurement of 0.03% to 0.01% when operated under controlled environmental conditions.

6.4 BOURDON TUBES, CAPSULES AND BELLOWS

Bourdon tube pressure gauge is constructed of a non-circular cross sectional tube formed into circular form. One end of this tube is fixed and the pressure is applied to the fixed end. The free end uncoils under action of pressure and a pointer attached to this end gives an indication of applied pressure along a graduated scale (fig. 6.3.A).

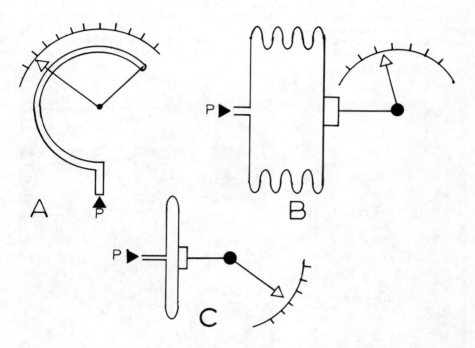

Fig. 6.3 Mechanical sensing elements: A. Bourdon tube; B. bellows; C. capsule.

There are a variety of shapes used in the construction of Bourdon tubes: coiled, spiral and helical. However, the most commonly used type is the 'C' shape which translates the free end movement into angular movement of a pointer by the use of a quadrant. The shape selected for a particular device depends on the measurement range and final production costs. The 'C' shape is generally used to measure up to 6 MPa and coiled shape is used above this range. The spiral measuring elements are normally employed for special applications.

Bourdon tube type pressure gauges are constructed to measure pressures in the range of 60 kPa to 1 GPa. A well designed pressure gauge of this type will have inherent temperature compensation and have a typical non-linearity of

0.1%.

The capsule (fig. 6.3.C) sometimes called aneroid, is constructed by joining two diaphragms around the periphery by a technique such as welding or brazing. It is generally used for measuring low pressures, up to 2.5 MPa. The diaphragms used to construct capsules may be flat or corrugated. Small deflexions of a flat diaphragm, less than half of its thickness, is linearly related to the applied pressure. However for higher deflexions the non-linearity of the flat diaphragm is improved by the incorporation of corrugations. Pressure gauges based on diaphragm type construction or capsules have better temperature characteristics than the Bourdon tube types. One common use for the capsule, sealed with vacuum, in in aneroid barometers as the pressure transducer for measuring atmospheric pressure.

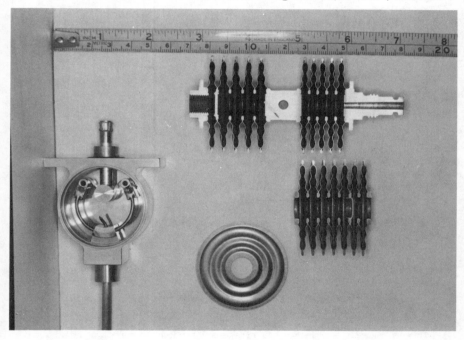

Fig. 6.4 Examples of commercially produced Bourdon tube device (left), single corrugated diaphragm and capsule stacks (Courtesy Negretti and Zambra (Aviation)).

The capsules may be stacked in order to obtain increased deflexion.

The bellows are constructed from thin walled tubing, having deep convolutions and sealed at one end which displaces axially when pressure is applied to the fixed end (fig. 6.3.B). They are generally used as a flexible pressure seal rather than pressure sensitive elements. When used as a pressure sensor, they exhibit good linearity and have a measurement range of 0.6 kPa to 100 kPa. However the stability of the zero position and the stiffness is not as good as other well designed mechanical sensors. It is usual to incorporate a good quality spring of higher stiffness factor in parallel with the pressure measuring bellows.

Examples of a commercially produced Bourdon tube device and stacked capsule assemblies are shown in fig. 6.4.

6.5 PRESSURE TRANSDUCERS

A pressure transducer consists of a force sensing device, usually a mechanical arrangement such as a diaphragm, whose deflexion under the applied pressure, is translated into an electrical signal by a secondary transducer. The advantage of such pressure transducers over the previously discussed mechanical devices is that they need very small deflexion of the force bearing member to produce a useful electrical output. They are usually small and light and have superior frequency response characteristic. In general, the accuracy of pressure transducers is higher than their mechanical counterparts, mainly due to the smaller deflexions allowed for the mechanical sensor. They lend themselves readily to cost effective production methods. The manufacture of certain types of pressure transducers, such as the piezoresistive type, benefits from the high technology developed for the manufacture of Integrated Circuits. Many of the applications, once handled by the traditional pressure sensors are now handled by the electrical pressure transducers which complies easily with the requirements of the modern process control systems of today's advanced process plants.

Most of these transducers can also be produced for use in two-wire systems where the output is in the form of a 4-20 mA current change.

However it must be remembered that these pressure transducers are not self indicating instruments and need additional electronics to produce a useful signal.

6.6 CAPACITIVE TYPE PRESSURE TRANSDUCER

This transducer has become very popular in the last decade especially within the automotive industry. There are two designs generally used: 1. single electrode system where the pressure is applied upon a diaphragm which moves with respect to the stationary electrode, 2. dual electrode system where the pressure bearing diaphragm is placed in the middle of the two fixed electrodes and allowed to move towards one electrode and away from the other. The change of capacitance between the diaphragm and the fixed electrode is a measure of the pressure applied. The capacitance is measured either by making it part of an oscillator whose frequency is modulated by the capacitance change or incorporating it into the Wheatstone bridge configuration. Fig. 6.5 illustrates the latter method.

Capacitive transducers exhibit inherent temperature

sensitivity and susceptibility to vibration and shock, however there are a number of designs to overcome these effects.

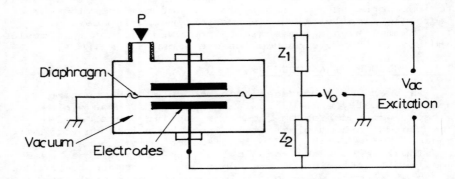

Fig. 6.5   Dual electrode pressure transducer and the detector circuit.

Modern capacitive pressure transducers are produced with a sputtered film single electrode on a ceramic substrate, employing a highly stable diaphragm with integrated electronics to convert frequency to voltage output. They can measure pressures from 0.1 kPa up to 1 GPa for special applications. The general purpose versions have a non-linearity of 0.5% with a temperature coefficient of sensitivity of 400 $PPMK^{-1}$ and the specially selected devices have accuracy of 0.05% with a comparable temperature coefficient of 20 $PPMK^{-1}$.

6.7   RELUCTIVE TYPE PRESSURE TRANSDUCER

There are two main types of pressure transducers using reluctive elements as the secondary transducer. These are the LVDT type and the inductance type. The former employs Bourdon tubes, bellows or capsules as primary sensing element (see 7.7 for operating principles of LVDT). Fig. 6.6.A illustrates a pressure transducer of this type where the deflexion of a capsule is transmitted via the core rod to the core of the LVDT which in turn translates this movement into an electrical signal. It is usual for these pressure transducers to incorporate DC-LVDT in order to minimise the external circuitry needed for its operation.

The inductance type utilises a diaphragm as the primary sensing element. The deflexion of the diaphragm is used to change the inductance of an electrical circuit (fig. 6.6.B). A pressure bearing magnetically permeable diaphragm is placed between two coils which are formed into an inductance bridge. The bridge output voltage changes as one inductance increases and the other decreases. Many manufacturers producing this type of pressure transducer incorporate the electronics into the transducer so that they can be excited from d.c. power lines and provide d.c.

output.

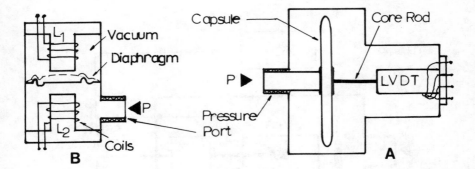

Fig. 6.6 Schematic illustration of A. LVDT type pressure transducer; B. reluctive type pressure transducer.

The measurement range can be up to 35 MPa with an accuracy of 0.1% for a well designed transducer used in favourable application conditions.

6.8 FORCE BALANCE PRESSURE TRANSDUCER

The operating principle of this technique is explained in 7.4. Fig. 6.7 illustrates a pressure transducer using a diaphragm as the pressure sensor. The deflexion of the diaphragm is sensed by the LVDT displacement transducer whose output is amplified and used to drive the servo actuator in order to restore the diaphragm to its original position. The current, i, flowing in this servo loop is a measure of force needed to restore the diaphragm to its original position. The voltage drop across the sensing resistor R_s is used to indicate the applied pressure.

The measurement range of these devices can be up to 500 kPa with an accuracy of 0.05% and repeatability of 0.02%.

6.9 PIEZOELECTRIC PRESSURE TRANSDUCER

These transducers are used in dynamic measurement applications where high frequency response, up to 500 kHz, is required [see also 7.3.4]. A piezoelectric crystal such as quartz produces no charge when subjected to hydrostatic pressure. However a charge output is produced when a force is applied to this crystal by means of a force bearing diaphragm inducing mechanical stress throughout its body.

A typical piezoelectric transducer is constructed from a stack of quartz crystal disks which are mechanically pre-loaded between two metallic electrodes. The design may also include a charge amplifier within the transducer housing to provide a low impedance output. This is a very useful

addition since the output impedance of piezoelectric
crystal devices are usually very high, typically 100 T ohm.

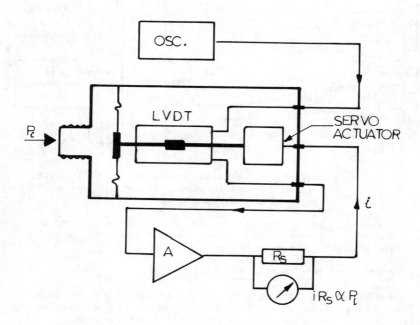

Fig. 6.7 Pressure transducer utilising force balance technique.

The pressure measurement range is usually up to 150 MPa although for short transient measurement of up to 1.5 TPa can be obtained with the use of specially constructed devices using lithium niobate crystal. The limiting temperature of operation for the crystal is the Curie point temperature which is 573°C for quartz and 350°C for most ceramic types. However the operating temperature range for the total transducer is limited by its construction, usually, from 200°C down to cryogenic temperatures.

6.10 STRAIN GAUGE PRESSURE TRANSDUCER

This is perhaps the most popular device for measuring pressure in industry. Unbonded wire strain gauges were first used in their construction. It is now common to use bonded foil or bar semiconductor and thin film deposited strain gauges. The bonded semiconductor strain gauge technique is now almost completely replaced by integrally diffused strain gauges where a silicon wafer diaphragm which is also used as the pressure sensor, is diffused directly with donor elements to obtain strain gauges, sometimes called piezoresistive elements, at defined locations.

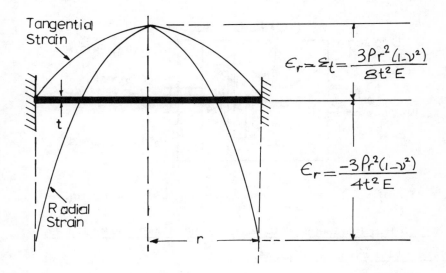

Fig. 6.8 Strain distribution in a clamped diaphragm.

The basic operating principles of strain gauge and the Wheatstone bridge is described in section 7.8. The diaphragm is popularly used as the pressure bearing member and allowed to produce strain fields (fig. 6.8). The tangential strain, ε_t, and radial strain, ε_r, are sensed by strain gauges located on these areas. A diaphragm strain gauge (fig. 7.8.B) designed for a specific diaphragm diameter, comprises 4 strain gauges, two of which are positioned on the tangential strain field near the centre and two positioned on the radial strain field near the edge. All strain gauges, semiconductor (piezoelectric) or diffused types mounted directly on diaphragms are positioned to detect the above strain levels.

Another type of pressure transducer makes use of a strain gauge force transducer as a displacement sensor to sense the deflexion of a diaphragm to produce an electrical output (fig. 6.9). A connecting rod transmits the deflexion to the force transducer, usually bending beam type with stiffness much lower than the diaphragm, and may have metal foil vacuum deposited thin film strain gauges connected in a Wheatstone bridge configuration.

A typical electrical circuit diagram for a strain guage pressure transducer is the same as for a load cell (fig. 7.10) with the same compensation and calibration resistors.

In this group, a wide variety of pressure transducers are produced to cover measurement ranges from 20 kPa to 250 MPa. A typical transducer will have an accuracy of 0.25 - 0.1% and have a temperature coefficient of span and zero of 50-100 PPMK^{-1} in the operating range of -20°C to +50°C.

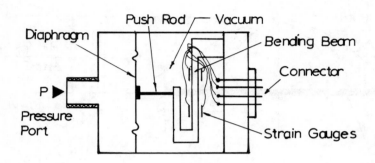

Fig. 6.9 Pressure transducer employing bending beam type load cell.

6.11 OTHER PRESSURE MEASURING METHODS AND TRANSDUCERS

6.11.1 Vibrating Wire Type.

As described in section 7.6, the change of resonant frequency of a vibrating wire or plate, tube, etc. as a function of force applied to it can be exploited to construct pressure transducers. A typical device employs a pressure sensing diaphragm, the centre of which is attached to a thin, taut wire. The wire is located in a magnetic field and a current is passed through it. The emf induced due to this current is detected, amplified and fed back to the wire to sustain its oscillation. The frequency of this oscillation is a measure of the force applied to the wire or the pressure applied to the diaphragm.

These transducers are used in aerospace, oceanography and civil engineering applications, and can have excellent repeatability. A well designed transducer will have built in linearity and temperature compensation circuitry to achieve up to 0.04% accuracy.

6.11.2 Potentiometric Pressure Transducer.

This is one of the earliest pressure transducers developed. It comprises, basically, a resistive potentiometer driven by a mechanical pressure sensor such as a capsule or Bourdon tube. Their output varies between 0 and 100% of the applied excitation voltage and they are inherently high level output transducers and can drive an indicator without the need of complicated electronics.

A capsule is used for the low range devices, for the high ranges Bourdon tube is employed in conjunction with a single or multi-track potentiometer. In some transducers nonlinear tracks are used to compensate the nonlinear nature of the mechanical sensors or to provide a linear output with nonlinear change of pressure such as in altitude meters.

The commercially available transducers of this type will have a range of 100 kPa to 50 MPa.

6.11.3 Resistive Pressure Transducer.

Certain conductive materials change their electrical resistance when subjected to hydrostatic pressure. This property has been utilised to construct resistive pressure transducers whose output resistance change is directly related to the pressure. The suitable materials are carbon, zirconium tetrachloride and manganin. The latter, an alloy of Cu, Mn, and Ni is the most commercially used material and it is manufactured in the form of a bondable strain gauge pattern (fig. 7.3.E), manganin gauge, and usually produced by strain gauge manufacturers. Manganin gauges change their resistance linearly with pressure and have a typical sensitivity 0.0027 ohm/ohm/100 MPa. They are used to measure very high pressures, up to 1.4 GPa and to study high pressure shock waves up to 40 TPa since their response time can be as low as a few nanoseconds.

6.11.4 Novel Pressure Transducers and Pressure Measurement Techniques.

There are a number of novel pressure transducers that have been either cited in literature, or used for research purposes or only available commercially for special applications. A selection of these are listed below:
1. Use of Hall effect devices to measure deflexion of a pressure bearing diaphragm,
2. sensing of eddy current losses in deflecting diaphragms,
3. driving of angular encoder by Bourdon tube to produce binary or BCD outputs,
4. Bourdon tube made from fused quartz,
5. use of hydrostatic pressure sensitive properties of planer transistor and a number of other crystals such as iridium antimonide,
6. a variety of micromechanical devices based on the pressure sensitive property of silicon monolithic IC's,
7. resonating quartz sensor.

BIBLIOGRAPHY

Angell, J.B., Terry, S.C., and Barth, P.W., 1983, 'Silicon Mechanical Devices', Scientific American, 248, No.4.

Julien, H., 1981, 'Handbook of Pressure Measurement with Resilient Elements', Alexander Wiegand, Main, West Germany.

Liptak, B.G., (Editor), 1969, 'Instrument Engineers' Handbook', Chilton Book Co., Philadelphia:New York:London.

Mitchell, A.R., 1981, 'Pressure and Vacuum Standards Review', Transducer Technology, 4, issue 2, p9 and issue 4, p8.

Neubert, H.K.P., 1975 'Instrument Transducers', Clarendon Press, Oxford, England.

Norton, Harry N., 1982 'Sensor and Analyser Handbook', Prentice-Hall Inc., Englewood Cliffs, N.J., U.S.A.

Chapter 7

Force and weight measurement

U. Erdem

7.1 INTRODUCTION

Force is the primary input for many of the physical quantities such as weight, pressure, acceleration, torque, stress, altitude, thrust and sound, and all are essential to engineering measurement. Force is a measure of attraction or repulsion between masses and has a unit of newton, N, in SI system of units. It is defined as that force acting on a mass of one kilogram that produces an acceleration of 1 ms^{-2}. Mass is a fundamental quantity and it has the unit of kilogram that is equal to the mass of the international prototype of the kilogram which is a solid cylinder of platinum-iridium kept by the International Bureau of Weights and Measures in Sevres, France.

Force and mass are related through Newton's Law of Motion which gives the relationship between an acceleration of a acting on a mass of m as $F=m.a$.

Weight is the force produced by the attraction between the mass of earth and mass of a body. The acceleration produced by this attraction is the gravitational acceleration, denoted as g. The actual value of g depends on the geological location and the distance between the earth and the mass to be measured.

The standard value of g is 9.806 ms^{-2}, however this varies about 0.1% in the U.K. and as much as 0.5% between the equator and the poles.

The measurement of weight is a special case of force measurement and it is highly developed directly due to it being the principle means of determining the quantity of goods exchanged or sold. Weighing is also important for efficient management of modern industrial plant employing material handling and storage. Most of the industrial weighing systems employ a force transducer known as a load cell which can be electrical, hydraulic or pneumatic.

Modern weighing instrumentation designed for use in control system loops such as batching, blending and discharging utilise, in majority of the applications, electrical load cells which are strain gauge based force transducers, usually known as electronic load cells.

Since the first utilisation of strain gauge load cells in early 1950's, much work has been done to improve the accuracy and long term stability of these transducers. It is now possible to design and install a complex weighing system and obtain total system accuracy of 0.02% of weigh range and does not require recalibration for up to two years.

As modern process control systems increasingly employ electronic signals for their data processing, only a limited reference will be made to other weighing methods and force transducers operating on hydraulic, pneumatic or mechanical principles.

There are basically two methods of weight measurement. These are direct comparison method and indirect comparison method.

7.2 DIRECT COMPARISON METHOD

This method employs a beam balance of equal arms utilising a null balance technique. The apparatus is generally known as the analytical balance. The movement produced by the unknown mass M_x is directly compared with the movement of the standard mass M_s. In the null position, indicating the beam is horizontal, the two masses or forces produced by them are equal if the arms are equal (fig. 7.1).

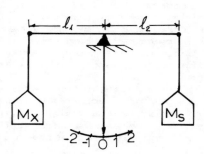

Fig. 7.1 Equal arm beam balance.
When horizontal $M_x = M_s$ for $l_1 = l_2$.

These balances are normally used for scientific purposes and are not generally suitable for industrial use. This type of apparatus can be used for comparing masses up to 1000 kg with a 1 PPM sensitivity. Specially constructed equal arm balances operated in controlled environments can achieve sensitivities better than 0.001 PPM.

7.3 INDIRECT COMPARISON METHOD

The limitation of direct comparison method becomes apparent when large weights have to be measured. It becomes necessary to employ beams of differing arm lengths or multiple lever systems to obtain reduction of the larger weights so that they can be counterbalanced by smaller and more manageable weights. Commercially available, and still popularly used, steelyard based weighing set-ups such as platforms and weighbridges are examples of this. Another method of measuring weight is by the use of calibrated transducers which produces an output suitable for control or display applications. This may be in the form of a pressure from a pneumatic or hydraulic load cell, or an electrical signal from an electronic load cell.

The lever systems, single or multiple, and transducers will have to be calibrated against known masses using Dead Weight force generators or with the use of Transfer Standard Transducers, such as proving rings and electronic load cells which are already calibrated by the use of Dead Weight force generators.

The individual weights in a Dead Weight machine are adjusted against a mass produced by the NPL to give the required force in the specific location where the machine is to be installed. This operation requires the use of a highly sensitive equal arm beam balance and the exact value of the gravitation for the site.

The Dead Weight machine, with an accuracy of 1 part in 50000 used for calibration of precision electronic load cells and proving rings, employed by Negretti and Zambra Ltd., on their Stocklake site is located at,

Latitude : 51° 48' 56" N
Longitude : 0° 47' 39" W
Altitude : 78.15m (Newlyn Datum)

The acceleration of free fall for this site is

$g = 9.81207 \text{ ms}^{-2}$

The calibration methods and accuracy classifications of these machines and transfer standard devices are governed by the British Standard 1610.

The National Weights and Measures Laboratories has the responsibility to maintain and disseminate the mass standards in the U.K. and the National Physical Laboratories has the same responsibility for the force standards in the U.K.

7.4 FORCE BALANCE TECHNIQUE

There is a diverse range of apparatus operating on the force balance principle to measure weight and pressure. It is unique in its operation in that a closed feedback loop is used to compare the electrical output to the force input. The deflexion produced by the applied force is sensed by a displacement transducer, output of which is amplified and employed to produce a restoring force equal to the applied force. Fig. 7.2 illustrates a weighing apparatus in which the deflexion of the load bearing spring caused by the weight placed on the weigh pan is sensed by the displacement transducer whose output is amplified and used to drive the force motor to restore the spring to its original position. The current i flowing in this feedback loop is a measure of the weight placed in the pan.

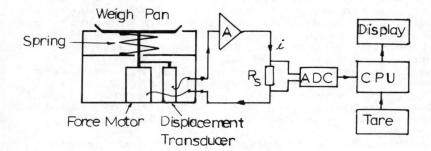

Fig. 7.2 Schematic illustration of a force balance weighing apparatus.

Because of the practical limitations in restoring high force levels, these instruments are generally limited to measuring forces below 1 kN. A typical weighing instrument of this type will have a measuring range of up to 20 kg with a resolution of 0.1 microgram to 0.1 gram. The advantage of this sytem is that is has a null deflexion resulting in a superior dynamic performance. It is also independent of zero instability, hysteresis and temperature effects. Due to its high accuracy and stability, these instruments are commonly used for counting components in stores and it generally competes with the analytical balance.

7.5 HYDRAULIC AND PNEUMATIC LOAD CELLS

It is possible to measure force by transducing it into pressure by the use of hydraulic or pneumatic load cells. These devices work on the principle that when a force is applied to one side of a piston or diaphragm then there has to be an equal and opposite pressure on the other side to balance the applied force. This pressure is a measure of the applied force and can be indicated by the use of a pressure dial having a sensing system such as a Bourdon gauge, or translated into an electrical signal by an electrical pressure transducer.

An example of a hydraulic load cell designed to measure forces in steel rolling mills is given in fig. 7.3.

These devices are self contained and do not need external power, this makes them ideal for portable systems and for hazardous area applications. Weighing system accuracies of 0.5 - 1% is achievable with careful design.

Pneumatic load cells are constructed somewhat similar to the hydraulic load cells, however, they employ a diaphragm or a flexible material to balance the applied force rather than pistons commonly used in hydraulic load cells.

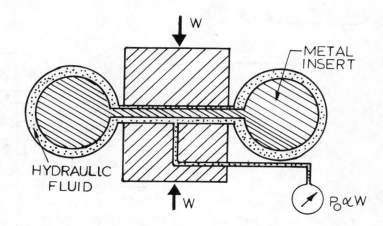

Fig. 7.3 Schematic diagram of a compression type toridal hydraulic load cell.

A typical pneumatic load cell shown in fig. 7.4 incorporates a bleed valve whose position is controlled by the deflexion of the diaphragm. There are a variety of pneumatic load cells commercially available with refinements and additions. The weighing system accuracies are similar to those of hydraulic load cell based systems.

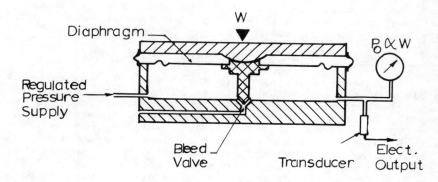

Fig. 7.4 Schematic diagram of a pneumatic load cell.

7.6 VIBRATING WIRE FORCE TRANSDUCER

The resonant frequency of a tensioned wire is a function of the tension applied to it. This principle is exploited to construct vibrating wire load cells and pressure transducers. It is basically a digital transducer with a frequency output as a function of applied force.

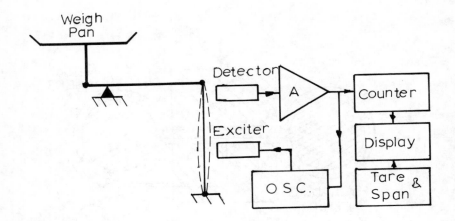

Fig. 7.5 Vibrating wire weighing apparatus.

Fig. 7.5 illustrates the use of a vibrating wire force transducer in a weighing apparatus. The resonant frequency of the wire is picked off with a detector head, may be counted and displayed directly without the use of Analogue to Digital Converter (ADC). The relationship, however, between the resonant frequency and the tension in the wire if of a nonlinear nature, around 0.5% and needs to be linearised for accurate measurements. A well designed transducer incorporates some means of inherent nonlinearity and temperature compensation techniques. Such transducers can then be accurate to 0.05%. The load bearing capacity is normally limited to 15 - 20 kg thus necessitating the use of lever systems to measure larger forces.

7.7 LINEAR VARIABLE DIFFERENTIAL TRANSFORMER

This transducer, commonly known as LVDT, is basically a displacement transducer. It is utilised to give an electrical output as a function of the deflexion of force bearing members in load cells and pressure transducers. The LVDT is essentially a specially designed transformer with a movable core and consists of a primary and two identical and symmetrically placed secondary windings that are magnetically coupled with the movable core (fig. 7.6).

The primary winding is energised with an a.c. reference voltage, the secondary windings are connected in opposition so that the combined output is zero when the core is in the central position. Any displacement from this null position results in an increase of output with a phase difference of 180° on each side of the null position. The amplitude of this output is proportional to the displacement input.

The LVDT's generally do not suffer from hysteresis problems due to its core having no contact with the main assembly.

Force and weight measurement   111

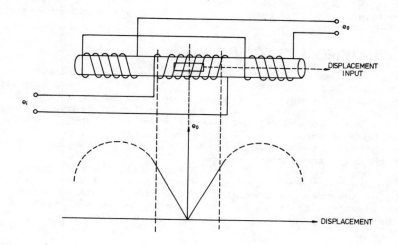

Fig. 7.6   Schematic diagram of LVDT and its
           input-output relationship.

The use of LVDT to sense deflexion of a hexagonal shaped proving ring as part of a load cell is shown in fig. 7.7.

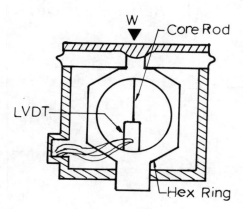

Fig. 7.7   LVDT type load cell.

The typical nonlinearity and repeatability of LVDT based transducers are 0.1% to 0.05% with a typical sensitivity of 2mV/0.25mm/V of reference input at excitation frequencies of 50 Hz to 20 kHz. DC-LVDT devices are also available incorporating microelectronics for the necessary conversion. These devices will work from typically ± 15 Vdc supply lines and have slightly degraded performance specifications compared to their a.c. operating counterparts.

7.8 STRAIN GAUGE FORCE TRANSDUCER

A strain gauge based force transducer can be used in weighing applications as a load cell or can be incorporated into a pressure sensing device and be used as pressure transducer. It basically consists of a force bearing elastic member to which a number of strain gauges, usually four to form a Wheatstone bridge, are bonded. The deflexion of the elastic member is translated into resistance change through the strain gauges.

The operating principle of a resistance strain gauge is based on the fact that when a piece of wire is stretched its resistance changes as a result of its dimensional changes. The resistance R or a wire length L, cross sectional area of A and resistivity ρ is given by the expression,

$$R = \frac{\rho L}{A} \qquad \ldots\ldots [7.1]$$

when the wire is strained, each of the above quantities will cahnge, therefore, partial differentiating equation 7.1 for a circular cross section of diameter D we obtain,

$$\frac{dR/R}{dL/L} = 1 - 2\frac{dD/D}{dL/L} + \frac{d\rho/\rho}{dL/L} \quad \ldots [7.2]$$

but,

$$\frac{dL}{L} = \epsilon_a \qquad \text{axial strain . [7.3]}$$

$$\frac{dD}{D} = \epsilon_t \qquad \text{transverse strain} \ldots [7.4]$$

$$\frac{dD/D}{dL/L} = \frac{\epsilon_a}{\epsilon_t} = \nu \qquad \text{Poissons ratio} \ldots [7.5]$$

$$\frac{dR/R}{dL/L} = F \quad \text{gauge factor} \quad \ldots \ldots \quad [7.6]$$

rewriting the equation 7.2 noting that axial and transverse strains have opposite signs.

$$F = 1 + 2\nu + \frac{d\rho/\rho}{\epsilon_a} \quad \ldots \ldots \quad [7.7]$$

the last term in this equation is the piezoresistance effect which can be positive or negative, the Poisson ratio is always positive and for most materials is less than 0.5.

In practice the strain gauge manufacturer specifies the gauge factor and the resistance value, rewriting equation 7.6,

$$\Delta R = FR\epsilon_a \quad \ldots \ldots \quad [7.8]$$

this expression gives the resistance change of a given strain gauge when subjected to a strain of ϵ_a. As an example, a typical commercial load cell working on 1000 μm/m strain level using 120 ohm strain gauge with a gauge factor of 2 will have a resistance change of 0.24 ohm.

The gauge factor for commonly used materials such as constantan and nickel-chrome alloys is 2. However there are other alloys used in the manufacture of strain gauges that exhibit gauge factors higher than 2. Examples are platinum-tungsten alloys 4.5 and isoelastic alloy 3.5

There are a variety of strain gauges for various applications, the most commonly used and commercially available, is the metallic foil type. The others are semi-conductor, vacuum deposited thin film, wire, unbonded wire and thick film. A number of foil strain gauge patterns are shown in fig. 7.8.

The strain level in an elastic force bearing member is a function of its geometrical shape and the modulus of elasticity of the material as well as the force applied. The shape selected for a particular device will depend on factors such as the required measurement range, final performance specification and cost of production.

The materials used for the force bearing elastic member are usually tool steels, aluminium, beryllium copper and precipitation hardened stainless steels. The basic requirement from the material is a linear stress-strain relationship with low hysteresis and creep specification. It is usual to heat treat these materials to achieve the above requirements as well as long term dimensional stability.

Cross sectional view of a sealed industrial load cell is shown in fig. 7.9. A typical commercially produced industrial load cell using foil strain gauges will have an accuracy of 0.05% with zero and span temperature co-efficients of 20 PPM K^{-1} and an output of 1 to 4 mV/V of excitation voltage at their rated load input. The circuit diagram of a Wheatstone bridge incorporating the compensation and calibration elements is shown in fig. 7.10.

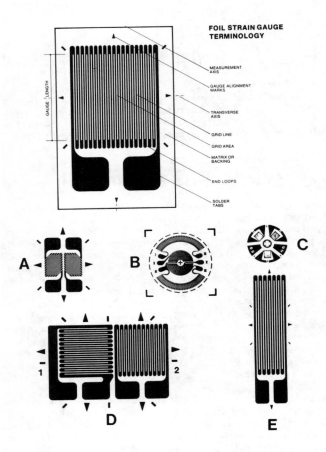

Fig. 7.8 Foil strain gauge patterns: A. two element 90° shear gauge; B. four element Wheatstone bridge diaphragm gauge for pressure transducers; C. three element 60° gauge; D. two element 90° gauge; E. single element gauge (Courtesy Micromeasurements.)

The load cells are produced in a wide variety of mechanical sizes and capable of measuring tension or compression loads from a few kg to several thousand tonnes. They are the most commonly used transducers in industrial weighing applications. Fig. 7.11 shows an industrial load cell mounted in a mechanical assembly to facilitate site mounting and prevent damage to the transducer by non axial loads which may be present.

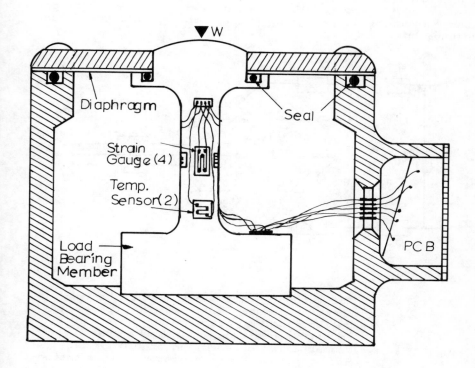

Fig. 7.9 An industrial load cell with a column load bearing member. (Courtesy Negretti and Zambra.)

7.9 OTHER WEIGHING METHODS AND FORCE TRANSDUCERS

7.9.1 Bouyancy Principle.

When an object is immersed in liquid it is lifted by a force equal to the weight of the liquid it displaces. An apparatus can be constructed to exploit this principle for weighing. However, this form of weighing is rarely used and most common use of this principle is in the estimation of a ship's cargo by measuring the difference in Plimsoll lines and relating this to the weight of the cargo.

7.9.2 Ballistic Weighing.

If a mass is suddenly applied to a spring then its deflexion will be twice the deflexion under steady state conditions. This fact can be used to construct a weighing apparatus with a lockable platform mounted on springs. Load is applied to the platform and the maximum deflexion is measured. In the absence of damping the deflexion is a measure of the applied load. This type of system is seldom used for commercial weighing and is not suitable for force measurement.

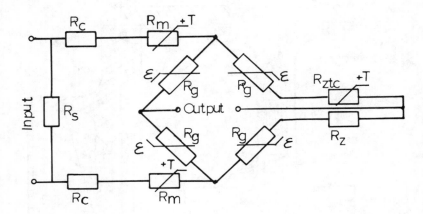

Fig. 7.10   Wheatstone bridge configuration for a force transducer. Rc resistor for span calibration; Rs resistor for balancing input; Rm resistor for temperature compensation of span; Rg strain gauge; Rztc resistor for temperature compensation of no load output; Rz bridge balancing resistor.

7.9.3 Gyroscopic Load Cell.

Force sensitive property of a gyroscope can be exploited to construct a load cell. If a load is applied to the outer gimbal and balanced by an equal and opposite load applied equally to the inner and outer gimbals, then the outer gimbal will rotate. The rate of precession of the outer gimbal is linearly related to the load applied. Commercially available gyroscopic load cells can be as accurate as 1 part in 100000. It is basically a fast responding digital force transducer. The device will bear loads up to 20 kg and higher loads will have to be reduced by a suitable lever system.

7.9.4 Piezoelectric Force Transducer.

Piezoelectricity occurs in crystals of certain configurations when subjected to tension or compression forces.

Most commonly used crystal is natural quartz, however, other natural crystals such as Rochelle salt and tourmaline can also be used. Piezoelectric properties can also be introduced into certain synthetic materials by polarising them. The transducers using these crystals as their transducing elements are considered as active devices since they generate charges and are used for dynamic force measurement applications. Their finite insulation resistance and measurement system loading on the output results in a finite steady state response for the transducer. These devices are commercially available and one common application is in the measurement of multiaxis forces in cutting machinery.

Fig. 7.11 A strain gauge based load cell in a side load protective mechanical assembly (Courtesy Negretti and Zambra).

7.9.5 Magnetoelastic Load Cell.

The operation of this load cell is based on the principle that when a load is applied to a ferromagnetic material, the magnetic movements of the Weiss domains change, resulting in a change of the magnetic properties of the material. A special transformer can be constructed with primary and secondary windings wound at right angles through diagonally opposite holes in the core which is made from a magnetically isotropic material. The primary is normally excited at mains frequency from stabilised supply lines and has an output impedance of 0.5 - 5 ohm.

This type of load cell is commercially available and

once compensated and linearised these devices can have an accuracy of 0.05%, and a repeatability of 0.02% in the operating temperature range of $-40^{\circ}C$ to $+100^{\circ}C$.

7.9.6 Novel Ideas.

There are a number of novel force transducers that have been cited in literature. A selection of these are listed below, however, at present, they may not be commercially available.
1. microwave cavity devices
2. surface acoustic wave oscillator devices
3. use of fibreoptic elements
4. ferromagnetic ribbons
5. use of conductive silicone rubbers
6. tunnel diode strain sensor devices

BIBLIOGRAPHY

Avril, J., 1974, 'Encyclopedie Vishay d'Analyse des Contraintes', Vishay Micromeasures, Malakoft, France.

Considine, D.M., 1974, 'Process Instruments and Controls Handbook', McGraw-Hill, New York, U.S.A.

Erdem, U., 1982, 'Force and Weight Measurement', J. Phys. E: Sci. Instrum., 15, 857-872.

Herceq, E.E., 1980, 'Handbook of Measurement and Control', Schaevitz Engineering, New Jersey, U.S.A.

Window, A.L., and Holister, G.S., 1982, 'Strain Gauge Technology', Applied Science Publishers, Barking, England.

Van Santen, G.W., 1967, 'Electronic Weighing and Process Control', Phillips Technical Library, Eindhoven, The Netherlands.

Chapter 8

Mathematical models and their use in force measuring instruments

Dr. F. Abdullah

8.1 INTRODUCTION

Mathematical modelling techniques are finding increasing applications in the analysis and design of instrument transducers and sensors used in measurement and control systems. Microprocessor based applications will lead to an increased requirement for sensors and actuators in the measurement and control field.

In the past the design of sensors and transducers has been approached in a semi-empirical method. Typically a physical transduction or sensing effect is discovered and then a large amount of experimental work is conducted in order to develop an instrument. The case of the orifice plate flowmeter illustrates this well. In principle the orifice plate is a simple sensing device; namely a circular plate with a hole in it. When inserted in a pipe a differential pressure is established either side of the plate. Experiments early in this century showed that the volume flow rate in the pipe was proportional to the square root of this pressure difference. The proportionality constant contains among other factors a quantity that needed to be determined experimentally namely the discharge coefficient. Following the feasibility of using the orifice plate as a flow sensing device a huge amount of experimental work (literally hundreds of man years of effort) went into establishing the discharge coefficient for different pipe and plate geometries, different upstream and downstream pressure tapping locations and different locations of the orifice in a pipe with respect to pipe bends.

The best modern industrial instruments are designed to have such low errors (typically force 0.02%, pressure 0.1%, displacement 0.1%, temperature 0.1% and flow 0.25%) that design and evaulation becomes a costly and time consuming exercise. The wide availability of cheap and powerful digital computers coupled with powerful numerical techniques has led to the possibility of developing accurate mathematical models to represent the static and dynamic response of typical instruments. A recent review paper (Abdullah and Finkelstein (1)) illustrates applications in the field of pressure, force, temperature, displacement and flow measurement.

Elastic elements are widely used as sensors for force and pressure measurement. Examples are Bourdon tubes, corrugated diaphragms and capsules, and bellows in pressure measuring insturments. In the area of force measurement elastic elements form the heart of many force measuring instruments. In weighing applications the elastic element is usually called a load cell billet and can take up a variety of forms. A convenient classification of these forms is according to the type of loading producing stresses in the elastic element. The common types are bending, compression/tension and shear billets. In this paper the mathematical modelling techniques that have been developed at the author's institution are introduced and illustrated in section 8.3 by way of application to strain gauged load cell transducers.

8.2 MATHEMATICAL MODELS AND THEIR USE

The aim of an instrument designer is to produce what is essentially an "information machine" which transforms the input to be measured into a convienent output (often but not always electrical) such that the output together with the transformation law can be used to deduce the input. In say an electrical linear pressure transducer the output would be a voltage v, the input a pressure p and the transformation law a straight line of slope k so that v=kp and p is determined from p=v/k.

The user of such an instrument would be quite happy with the instrument law or calibration. The designer would have to relate this to the subsystems and their manner of interconnection. Confronted with a requirement for a new design the designer may adopt one of two approaches (1) he may design using subsystems of known performance, (2) he may be forced to design the subsystems. To illustrate matters suppose a pressure transducer is to be designed using bellows. Then a typical system would be that the bellow under a pressure load is detected by some form of displacement sensor (a linear variable differential transformer, LVDT, perhaps) to produce an electrical output. A system maybe configured from available subsystems (bellows and LVDTs). At this level the designer may benefit from a mathematical model of the total instrument in terms of the subsystem characteristics; the subsystem characteristics having been obtained by experimental identification techniques or available from a subsystem manufacturer. This type of modelling has been called functional modelling [Finkelstein and Watts (2)]. In essence it is a generalisation of the electrical circuit modelling problem:- given a basic set of elements connected in a prescribed manner how does the total system respond to given inputs? The difference from conventional electrical circuit modelling is that we must be able to handle non electrical energies as well. The generalistion is achieved by identifying "through" and "across" (or so called "effort" and "flow") variables for each energy domain and defining a set of elements based on mathematical relationships between these variables. The elements themselves can be classified by the manner in which power is handled (either dissipated, stored, converted or transmitted). This leads to a basic set of one port (R,L,C) and two port (GY,TF) elements. R being a dissipative or resistive element, L and C the two types of storage elements and GY (Gyrator) and TF (Transformer) power conversion elements. These together with some purely signal transforming elements allows modelling of a wide variety of engineering systems and is particularly applicable to instrument transducers.

Once an instrument is described in some form of graphical representation showing the basic primitive elements and their manner of connection the differential equations describing the total instrument can be derived using techniques from circuit theory. The resulting time dependent equations could then be solved using one of a number of available computer packages. However, a better approach for the designer is to allow the computer to both <u>formulate</u> and <u>solve</u> the equations beginning from a suitable schematic or graph representing the system. This approach would enable the instrument designer, with little or no experience of differential equations or mathematical techniques, to develop models in a language that is familiar to him.

In the case of the pressure transducer illustration this would be in terms of the bellow's stiffness and effective mass (both storage elements and hence to be identified with a mechanical L and C) and the effective area (pure converter to be identified with TF or GY). In a similar way the mechanical and electrical elements of the LVDT displacement sensor could be readily identified. At the author's institution the above approach to modelling has been adpoted and an interactive package MEDIEM (<u>M</u>ulti <u>E</u>nergy <u>D</u>omain <u>I</u>nteractive <u>E</u>lement <u>M</u>odelling) has been developed to aid the efficient and rapid modelling of instrument transducers to obtain dynamic response information (both time and frequency responses). For further information on the "through and across" system modelling and the development and application of MEDIEM the interested reader is referred to reference (2) and a paper on MEDIEM [Liebner et al (3)]. An example of modelling of a complex differential pressure transducer comprising mechanical, fluid and electrical components using MEDIEM is illustrated in reference (3).

The other requirement that the instrument designer may have is to design the sensing and/or actuating subsystems of an instrument. Suppose no suitable bellows can be found for the pressure transducer illustration. In this case he will need to design the bellows in terms of its geometrical form and dimensions and the materials of which it is made. Mathematical models which would assist here are called <u>Physical</u> models(2). Such models require the calculation of "field" quantities which are obtained from the solution of appropriate partial differential equations. For elastic elements or devices this would be the displacement field; for electromagnetic devices - electric and/or magnetic fields; for flow velocity and pressure fields and for temperature sensors - heat flux and temperature fields. From such "fields" quantities of interest such as strains, stresses, stiffnesses, masses, resistances, inductances, discharge coefficients can be obtained by integration, differentiation or other simple operations.

A powerful method for the solution of appropriate partial differential equations which has become very popular in recent years is the finite element method which has largely superseded the finite difference method except perhaps in turbulent flow modelling. A number of excellent text books in finite element modelling are available, [Zienkiewicz, (4); Chari and Silvester (5); Taylor and Hughes (6)]. The trend in the development of finite element programs is towards the adoption of a modular scheme based on a library of subroutines specific to finite element manipulations. An example is the finite element library system adopted by the SERC [SERC (7)], which coupled with a computer with good graphics and editing facilities can lead to the efficient interactive development of complex finite element programs.

The main attraction of the finite element technique is that it can be used to model systems with complex geometrical configurations. This is because flexibility of element size, shape and distribution makes it possible to represent complicated boundaries. The example of the "square ring" billet modelling described in section 8.3 illustrates this well. Experience in the development and application of functional and physical models to instrument transducers and sensors has shown that physical models can be used to predict quantities of interest such as sensitivity and linearity errors to within a few per cent of experiment. On the other hand functional models can predict dynamic quantities of interest such as frequency response, overshoot, settling time etc. to within about 10-20%.

Our current capability in instrument modelling is illustrated in table 8.1.

Experience with the use of mathematical models in design studies indicates that modelling should proceed on a hiearchical basis with the adoption of simple and perhaps analytic models for initial design studies proceeding to more complex models as one approaches a final realisation. The case of shear force load cell billet design described briefly in section 8.3 illustrates this well.

Once mathematical models have been developed and validated by comparison with experiments they can be used to aid in the design, development and production stages of manufacture of an instrument. This is principally by employing sensitivity analysis techniques to determine the most important design variables or parameters which can

TABLE 8.1

Current Capabilities in Instrument Modelling

	INSTRUMENTS (generally with axi or plain symmetry in geometry and loadings)	TYPES OF PREDICTIONS
Elastic devices	• diaphragms, capsules, bellows, Bourdon tubes for pressure sensing • load cell billets for force sensing • snap action diaphragms	• sensitivity • linearity • displacements • stresses • strains • stiffness • effective areas and effective masses • Dynamic responses
Electro magnetic devices	• inductive and capacitive displacement sensors • fringe field electrode systems • magnetic circuits for transducers • torque motors and solenoid actuators	• sensitivity • linearity • saturation • air gap fluxes • inductance • resistance • capacitance • forces and torques
Flow devices	• turbulent flow through orifice plates, venturi and other head type flow meters	• Discharge coefficients • pressure loss
Thermal devices	• Platinum resistance thermometers and thermocouples in well systems	• time and frequency response • stem correction errors

then be used to explore alternative designs prior to experimental prototyping. Sensitivity analysis techniques can also provide valuable information for production tolerance purposes. In addition, dimensional analysis techniques are used to provide highly condensed design information from a minimal number of computer runs. This information may be displayed as a set of normalised design curves which can be used to develop a simple design methodology. The whole methodology of design using mathematical modelling techniques is illustrated well in a thesis on snap action diaphragms [Mirza (8)]. Such diaphragms are used in pressure and force activated switches or relays.

8.3 APPLICATION OF MATHEMATICAL MODELS TO FORCE TRANSDUCERS

In this section we indicate how mathematical models can be usefully used in force transducer design. The types of models range from a very simple use of mathematics to illustrate a concept, to the use of simple stress analysis where appropriate and finally to complex finite element models when other methods do not suffice.

Force transducers are generally divided into those used for weighing applications and those used for dynamic and/or multi component force measurement. The former are called load cells and used in essentially static conditions. They are required to respond to only one component of force namely that acting in the direction of the gravity vector at any location. The latter are complex and can be designed to measure multicomponents of the force vector. The force vector at a point will contain six components: three resolved forces and three resolved moments. Examples of applications of multi components force transducers are in sensors for robots and in sting balances used with aircraft models in wind tunnel testing. In dynamic applications the force transducer is often called a dynamometer.

We will consider the design of load cells as they are the simpler of the two types of transducer and yet illustrate design principles that are common to multi component dynamometers. Most load cell transducers are of the strain gauged type and these are considered here.

To illustrate the very simple use of mathematics we consider the input/output transformation law for a force transducer. Figure 8.1 shows a functional block diagram for a load cell

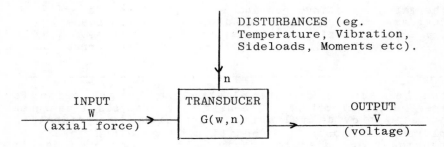

Fig. 8.1   Transducer Functional Block Diagram

The simplest transformation law is a linear one whence
$$G(W,n) = K \text{ (a constant)}$$
so that
$$V = KW \qquad (8.1)$$

A little thought leads to the conclusion that the above equation and its bounds highlight important design performance parameters.

Some of these are:-
(i) Sensitivity - the value of K.
(ii) Range - the values of W for which equation 8.1 is valid.
(iii) Linearity error - K is not constant in the range but is a function of W.
(iv) Hysteresis error - K at a particular W differs during the loading and unloading cycles.
(v) Creep errors - elastic after effect in materials implies equation 8.1 is not instantaneous.
(vi) Sideloads and bending moment load errors (resolved as two sideloads and three bending moment loads)
   - (a) span (or rated load) errors - K depends on such loads.
   - (b) zero (or no load) errors - $V \neq 0$ when W = 0 and such loads are applied.
(vii) Symmetry error - K differs for +ve (compressive and -ve (tensile) loads.
(viii) Temperature errors:
   (a) span errors - K depends on temperature change θ
   (b) zero errors - $V \neq 0$ when $\theta \neq 0$ and W = 0.
(ix) Dynamic errors:
   (a) Frequency - amplitude and phase errors
   (b) Time - % overshoot, settling time, and integral errors.
   (Note that creep, listed already, is a dynamic error).
(x) Port loading errors - K depends on mechanical input loadings and electrical output loadings.

In strain gauged load cells an elastic element (or billet) is strained by the load to be measured and the strains at particular locations on the billet are measured using electrical resistance strain gauges. Gauge resistance changes are detected in a Wheatstone bridge arrangement. The strain gauged load cell may therefore be conveniently divided into three subsystems (Fig 8.2).

The strain field $\underline{\epsilon}$ implies the value of all strain components at each point of the load cell billet. Of course the actual strain values required are those at the positions of the i strain gauges. The number of strain gauges can be from 1 to 4 or even more.

126 Mathematical models

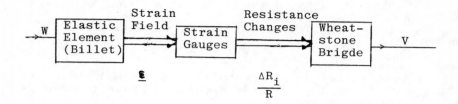

Fig. 8.2 Subsystems of the Transducer

Before proceeding to an analysis of the subsystems it is useful to list some of the performance requirements of a typical load cell:-

(i) "Universal" (tension-compression) operation
(ii) Linearity, hysteresis, repeatability, etc, better than 0.03% of Full Scale output (FSO), ie. better than 0.3 kgf for a 1000 kgf device.
(iii) Sensitive only to an axial load (ie. insensitive to the other resolved loads: 2 forces and 3 bending moments).
(iv) Insensitive to precise application point of the load (ie. insensitive to eccentric loading).
(v) Temperature "span" and "zero" compensated.
(vi) Full Scale Output of 2.0 mV/V Bridge input.

The meaning of some of these quantities will become more apparent as we analyse the design of a column type load cell (Fig 8.3).
In this type of load cell billet strain gauges are placed on the outer cylinderical surface of the cell at a location about halfway down. In this region there will be only one stress component, a compressive stress.

$$\sigma = \frac{W}{\pi(r_o^2 - r_i^2)} \qquad (8.2)$$

but two components of strain

$$\varepsilon_l = \frac{\sigma}{E} \quad \text{and} \quad \varepsilon_t = -\nu \varepsilon_l \qquad (8.3)$$

where ε_t is the transverse or θ direction strain due to Poisson's effect. E: Youngs modulus; ν Poisson ratio.
The strain sensed by a strain gauge R_i of nominal unstrained resistance R leads to an increased resistance:

$$R_i = R + \Delta R_i = R(1 + \Delta R_i/R) \qquad (8.4)$$

where

$$\frac{\Delta R_i}{R} = k \varepsilon_i \qquad (8.5)$$

ε_i :- strain for i_{th} gauge and
k :- gauge factor (assumed constant for all the gauges) Usually $k \approx 2$ for resistance strain gauges.

then $R_i = R(1+k\varepsilon_i)$ (8.6)

In strain gauged cells an unbalanced Wheatstone bridge (Fig. 8.4) is used to detect resistance changes in the gauges

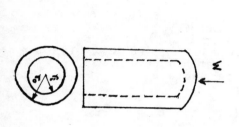

Fig. 8.3 A Column Billet   Fig. 8.4 Wheatstone Bridge

where V_s is the excitation voltage and V the output voltage. This output

$$V = V_A - V_B = \frac{R_1}{R_1+R_2}V_s - \frac{R_4}{R_3+R_4}V_s$$

$$\therefore \frac{V}{V_s} = \frac{R_1}{R_1+R_2} - \frac{R_4}{R_1+R_4} \quad (8.7)$$

put $R_i = R(1+k\varepsilon_i)$.   ($i = 1, 2, 3,$ or 4)   (8.8)

Then $\frac{V}{V_s} = \frac{1+k\varepsilon_1}{2+k(\varepsilon_1+\varepsilon_2)} - \frac{1+k\varepsilon_4}{2+k(\varepsilon_3+\varepsilon_4)}$ (8.9)

Equation (8.9) can be used simply, to illustrate the advantages of using more than one gauge. If only one gauge (say R_1) is used and all the others are fixed resistors of value R then (putting $\varepsilon_1 = \varepsilon$ and $\varepsilon_2 = \varepsilon_3 = \varepsilon_4 = 0$):-

$$\frac{V}{V_s} = \frac{1+k\varepsilon}{2+k\varepsilon} - \frac{1}{2} \quad (8.10)$$

Generally strain levels are kept small typically 1000 microstrain (1 microstrain - 10^{-6}) so $k\varepsilon \approx 2 \cdot 10^{-3}$

$$\frac{V}{V_s} = (1+k\varepsilon)(2+k\varepsilon)^{-1} - \frac{1}{2}$$

$$\simeq \frac{k\varepsilon}{4} \qquad (8.11)$$

ignoring non linearities.
Equation (8.9) shows that if four gauges are used with

$$\varepsilon_1 = \varepsilon_3 = \varepsilon \quad \text{and} \quad \varepsilon_2 = \varepsilon_4 = -\varepsilon \qquad (8.12)$$

then

$$\frac{V}{V_s} = k\varepsilon \quad \text{exactly} \qquad (8.13)$$

so that the output $\frac{V}{V_s}$ is increased by a factor of four and there is no strain-bridge non linearity.

When four gauges are used in a bridge the bridge is called fully active. When in addition (8.12) holds the bridge is called a fully active matched bridge.

Condition (8.12) cannot be achieved for the column billet the best possibility is

$$\varepsilon_1 = \varepsilon_3 = \varepsilon_L \quad \text{and} \quad \varepsilon_2 = \varepsilon_4 = -\nu \varepsilon_L$$

(two direct and two poisson gauges) so that

$$\frac{V}{V_s} = \frac{1+k\varepsilon_L}{2+k(1-\nu)\varepsilon_L} - \frac{1-k\varepsilon_L \nu}{2+k(1-\nu)\varepsilon_L}$$

$$\frac{V}{V_s} = \frac{k(1+\nu)\varepsilon_L}{2+k(1-\nu)\varepsilon_L} \qquad (8.14)$$

which is the design equation when

$$\varepsilon_L = \frac{W}{E\pi(r_o^2 - r_i^2)} \quad \text{is substituted.}$$

The equation will be non linear in W - this is the so called bridge induced non linearity.

Temperature induced errors should ideally be zero in a full bridge situation. This can be seen because a uniform temperature rise would produce the same strain in each gauge $\varepsilon_i = \varepsilon_\theta$ say and then $\frac{V}{V_s} = 0$ (from equation (8.9)).

Additional error analysis indicates that if sideload errors are to be minimised then Poisson and direct strain gauges should be equally spaced at 90° around the circumference of the cell (Abdullah et al (9)). The final arrangement is shown in figure 8.5.

A linear approximation to equation (8.14) leads to

$$\frac{V}{V_s} = \frac{(1+\nu)kW}{2E\pi(r_o^2 - r_i^2)} \qquad (8.15)$$

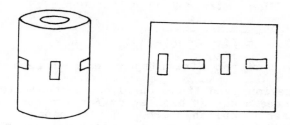

Developed Surface

Fig. 8.5 Arrangement of Gauges on Billet

This equation can be used for design. If a full scale output of 2mV/V is required then $V/V_s = 2 \times 10^{-3}$, W is known and E, ν are usually restricted. Therefore design is concerned with choosing r_o and r_i for the rated load W.

Even when r_o and r_i are chosen the design is not complete because the height of the load cell billet has not been specified. The height should be sufficient to ensure that a uniform strain field exists at the gauge locations. On the other hand it should not be excessivley tall because then stability problems may arise, also excessive material will be used.

A finite element model such as the one used for a "square ring" load cell billet to be described shortly could be used to optimise the height of the column load cell and also the loading button at the top of the the cell. Presently this is done by an experienced designer by "rules of thumb". For example the height would be a number of multiples of the outer radius r_o. A finite element model would be used to minimise the height while at the same time ensuring that the strains are uniform under the strain gauges and with values close to those predicted by the simple analysis.

We have shown that the bridge may introduce a non linearity in the column billet. To consider other sources of non linearity the block diagram of figure 8.2 is referred to. Here we can see that two other possible sources of non-linearity are from the load cell billet transformation (load to strain) and the strain gauge (strain to resistance changes); the latter because k in the strain gauge equation $\frac{\Delta R}{R} = k\varepsilon$ depends on strain. This non linearity, however, is very small and can be neglected. The load-strain non linearity however can be large particularly on column billets. This is because sectional area $A = \pi(r_o^2 - r_i^2)$ will change as the load is increased due to a Poisson effect. A point at a radial distance r will move to $r(1-\nu\varepsilon_l)$ so the area will change to $\pi(r_o^2 - r_i^2)(1-\nu\varepsilon_l)^2 \simeq \pi(r_o^2 - r_i^2)(1-2\nu\varepsilon_l)$

The non linear strain-load relationship is

$$\varepsilon_L = \frac{W}{E\pi(r_o^2 - r_i^2)(1 - 2\nu\varepsilon_L)}$$

This type of non linearity is called geometric non linearity

In a previous vacation school on instrumentation the author presented a design case study for a shear force load cell transducer (Abdullah (10)). In that paper simple beam theory was used to illustrate design principles and produce design equations for a 1000 kgf device. As in the column billet example some dimensions needed to be determined by more detailed finite element modelling. In the case of the shear cell a supporting thick ring was used to provied a clamped end conditions for the beam. The size of this was optimised using FE modelling.

In the case study an example of "gauge modelling" was provided. Gauge models are useful to calculate tolerances that can be allowed on the positioning of strain gauges on a billet to ensure linearity errors are small. The translation and rotational movement of a gauge about some ideal position often leads to a loss of output. If an analytic expression for the strain field in a load cell billet is available a gauge model can be developed to calculate strain levels for any size gauge located in non ideal positions on the billet.

We now consider finite element models of load cell billets. The analytic techniques described so far provide a method for designing truss (tension/compression) and bending beam like billets. The theory needs to be supplemented at times with finite element models to help produce a final realisation of a design. In other cases the FE technique allows modelling of billets where no possiblities exist of analytic modelling.

The basic idea behind the finite element method is that the continuum region to be modelled (in our case a load cell billet) can be approximated by replacing it with discrete elements assembled together. The unknowns or field variables (two components of displacement in our case) of an element are expressed in terms of interpolation functions within each element. The interpolation functions, also known as shape functions, are defined in terms of field variables at specific points called nodal points. Nodes usually lie on the element boundaries where elements are interconnected, although some elements may also have interior nodes. The behaviour of a field variable within an element is completely defined by the nodal values of the field variable and the shape functions for the element. Using finite element formulation, the nodal values of the field variables become the new unknowns. If these unknowns are found the shape functions define the field variables throughout the assemblage of elements. Once these fields variables are found other quantities of interest can be determined. For example, from the displacement field in a load cell the strain field can be calculated and this together with material properties (Young's modulus and Poisson's ratio) allows the stress field to be calculated.

In a recent thesis (Li (11)) a non linear finite element analysis program has been developed which can be applied to axisymmetric and plane symmetric load cell billet structrues. Compression colunm and tension rod billets are usually axisymmetric. Bending type billets such as the cantilever beam, proving ring, square ring, hexagonal ring, double ring etc. can be considered as plane stress structures. The application area of the program is therefore fairly wide.

The program was validated on comparison with other published theoretical and experimental work on proving ring billets. It was then applied to some commercial bending type billets and was used in the design of a "square ring" billet. The latter was designed, manufacured, gauged and tested in a final year BSc project extending over a few months. The present state of FE modelling for load cells has been summarised in a resent paper [Abdullah and Li (12)]. Here we present results for the square ring billet a half section of which is illustrated in Fig 8.6.

Symmetry of the billet shape and loading allows modelling of only half the actual billet. The model is divided into 126 eight noded quadrilateral elements each element having four corner and four midside nodes. There are 469 nodes on the entire region. In terms of finite element analysis the problem is of reasonably small size. The load cell was designed for a rated load of 50 kgf and tested with 30 lbf in tension. The approximate dimensions of the billet are 170 mm height, 96.6 mm width and 25 mm (into the plane of the paper). The circular hole is 94 mm diameter. As can be seen the density of meshing (elements) is increased where high stresses and strains are expected. Fig 8.7 shows a close up of the meshing in the region in which strain gauges are placed. Gauges are placed in the vertical direction on the outside and inside faces at the point of minimum thickness (1.3 mm).

Fig 8.8 shows a close up of the meshing in the other critical region. The displacement of the billet (fig 8.9) clearly shows that the billet is bending at the gauge locations. Fig 8.10 shows the strain contours at the gauge locations. Such a contour plot gives a clear indication of strain magnitudes and variations of strain in the gauging region and therefore is a very useful visual aid. A post processor called FEWVIEW for structural analysis was used to produce these results on a graphics VDU.

Table 8.2 below summarises the model and experimental results for this billet. A four active arm bridge measuring circuit was used in the experiments. In the model the gauge lengths were taken into account by averaging strains over each gauge.

Fig. 8.6   Half Section of a Square Ring Billet with FE Mesh

Fig. 8.8   FE Mesh in Corner Region of Billet

Fig. 8.7   FE Mesh for Square Ring Billet in Vicinity of Strain Gauges

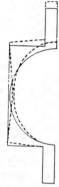 

Fig. 8.9 Displaced Shape of the Billet (Exaggerated 50 Times)

Fig. 8.10 Strain Contours in Vicinity of Strain Gauges

	Output V/V_s (mV/V)	Terminal Non Linearity Error (%)
Model	0.460	0.110
Experiment	0.503	0.119
Difference (%)	8	7.5

TABLE 8.2
Comparison of Finite Element Modelling with Experiment for a Square Ring Load Cell.

Results on a number of different load cell billets indicate that the agreement (better than 10% for both output sensitivity and non linearity error) is typical.

8.4 SUMMARY AND CONCLUSIONS

Strain gauged force transducers dominate the field of

load measuring techniques. In weighing the best modern load cells have errors of less than 0.1% and to achieve this degree of perfection the load cell billet, gauging and measuring circuit must be carefully designed considering all sources of errors.

Here we have attempted to show, in a tutorial manner, that mathematical models can assist in this task.

Mathematical models can be applied at various levels. To summarise:-

(i) Simple mathematics to identify the instrument transformation law and block diagram decomposition to reveal subsystems and their own transformations.
(ii) Analytic models based on truss and beam theory for simple load cell billet structrues.
(iii) Simple mathematics applied to the Wheatstone bridge measuring circuit to show how outputs can be calculated and errors are compensated.
(iv) Gauge models to produce output variations as gauges of a particular size are moved over a strained region.
(v) Finite element models that remove many of the restrictions (eg. simple geometries and loadings) of analytic models and can be used, when in doubt, to test the validity of the simpler analytic models.

Finite element analysis of the sort described here can now be performed on 8 bit and 16 bit microcomputer systems costing less than £10,000. With interactive graphics for display of strain fields, designs can be studied in great detail. We may therefore be sure that mathematical models are here to stay in force transducer design.

The techniuqes and methodology are applicable to the more complex case of design of multi component force transducers (eg. in robotics). In particular complex elastic structures can be studied using FE methods including cross axis couplings that arise when one component of the load vector can change the elastic behaviour of the structure for another component.

REFERENCES

1. Abdullah, F., and Finkelstein, L., 1982, "Review of Mathematical Modelling of Instrument Transducers' Acta IMEKO separatum pp. 145-157, Budapest, Hungary.

2. Finkelstein, L. and Watts, R. D., 1978, "Mathematical Models of Instruments: Fundamental Principles", J. Phys. E.: Scientific Instruments, Vol 10 pp 566-72.

3. Liebner, R. D., Abdullah, F., and Finkelstein, L., 1982, "Structure Graphs: A New Approach to Interactive Computer Modelling of Multi-Energy Domain Systems" J. Dynamic Systems, Measurement and Control Vol 104 pp 143-50.

4. Zienkiewicz, O. C., 1975, "The Finite Element Method in Engineering Science", McGraw Hill 3rd Edition.

5. Chari, M. K. V. and Silvester, P. P., 1980, "Finite Elements in Electrical and Magnetic Field Problems", John Wiley and Sons.

6. Taylor, C. and Hughes, T. G., 1981, "Finite Element Programming of the Navier Stokes Equations", Pineridge Press.

7. SERC Level 0 and Level 1 Finite Element Library Documentation available for the Computer Applications Group, Rutherford and Appelton Laboratories, Oxford.

8. Mirza, M. K., 1983, "Mathematical Modelling and Design of Snap Action Diaphragms" PhD Thesis, Department of Systems Science, The City University. London, EC1V OHB

9. Abdullah, F., Erdem, U. and Mirza, M. K., 1980, "Linearity and Loading Errors in High Precision Load Cells", IMEKO Symposium published in "Weighing Technology", Krakow NOT pp 105-111, Poland.

10. Abdullah, F. 1979, Instrument Design Case Study: A 1 Tonnef Load Cell. Lecture given at SERC Vacation School in Instrumentation and Measurement, 2-6 July, 1979. The City University, London, EC1V OB4.

11. Li, C. W. 1981, "Non Linear Analysis of Load Cell Billets using the Finite Element Technique" MSc Thesis, Department of Systems Science, The City University, London, EC1V OHB.

12. Abdullah, F. and Li, C. W. 1983, "Finite Element Modelling of Load Cell Billets", presented at WEIGHTECH 83 conference and published in "Weighing and Force Measurement in Trade and Industry", Inst. M. C. publication.

Chapter 9

Liquid level measurement

B. Baldwin

9.1 INTRODUCTION

The measurement of liquid level is a fundamental one used in the automatic control of continuous processes. It is frequently used in conjunction with other basic measurements of temperature, pressure and flow for the control of processes in chemical and petroleum industries and is of prime importance in water works, power stations, steam raising plants and a number of other applications.

Several principles of measurement are used in determining the level of liquids. The type of instrument selected being governed by, the nature of the liquid, the shape of the vessel in which the liquid is contained, the pressure under which it is operating, and the application.

To enable the various instruments used to quantify the measurement made, various units are used. Linear units such as Metres for a direct measurement of depth or pressure units such as bars for a pressure head.

9.2 METHODS OF LEVEL MEASUREMENT

Level can be measured in a number of different ways. The simplicity or complexity of the instrument used will depend largely on the application of the measurement, whether it is an infrequent measurement made for long term records or a continuous measurement needed for the automatic control of a complex process.

The main types used in the process industries can be grouped under the following classifications.

1. Visual Indicators
2. Float actuated instruments
3. Displacement type instruments
4. Hydrostatic pressure instruments
5. Differential pressure instruments
6. Probe methods
7. Radio frequency methods

A description of these types, generally encountered in process and industrial applications follows.

9.2.1. Visual Indicators

The simplest, and probably the most common method of measuring level in an open tank, river or flume, is by means of a dipstick or gauge staff immersed in the liquid and marked off in contents or depth over a datum line.

The dipstick although crude and simple is a very accurate method of level measurement but cannot be used for automatic recording or controlling purposes. It has many applications where a continuous indication is unnecessary but where regular readings can easily be taken. A very common application of the dipstick is known by every motorist when he regularly checks his oil level.

A development of the simple dipstick is the hook gauge used where accurate measurement of the liquid head of a river or open tank is required and where it is difficult to align the eye with the liquid surface. This consists of a sharp pointed hook attached to a vernier scale mounted on a gauge staff (fig.9.1)

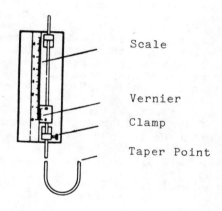

fig.9.1   Hook gauge

The hook is lowered into the liquid and gradually raised until the point of the hook just breaks the surface, allowing the level to be read off on the vernier scale.

An optical version of the hook gauge is found in the reflecting point manometer in which a steel point is placed pointing upwards in the liquid. This point is viewed through an eye piece at an angle in such a manner that an inverted reflection of the point is also seen in the eye piece (fig.9.2)

138 Liquid level measurement

Fig.9.2   Reflecting point manometer

The point is raised, and the level read, when the tips of the viewed and reflected points meet. This method of measurement, although elaborate, overcomes surface tension problems encountered when trying to estimate when the tip of a standard hook gauge breaks the liquid surface.

Another type of visual indicator is the sight glass, consisting of a transparent tube mounted on the side of a vessel and connected to it by pipes at the top and bottom. The liquid in the tube rises to the same level as in the tank and its height can be compared and recorded against a graduated scale behind it. Sight glasses are frequently used to measure the water level in the drum of a boiler. Such devices, because of the high pressures involved, are constructed within a steel chamber with a thick glass opening front and back. By the use of a two coloured glass strip behind the sight glass and the refractive effect of water, a clear indication of the water level is obtainable, the space above the water line appearing blue and that below the water line, red.

These methods of level measurement require easy access to the liquid surface. If, however, the tank is elevated away from an accessible position, a means of transmitting the measurement to a more convenient position needs to be applied. Such a device is to be found in the balanced float method.

A float resting on the surface of the liquid is connected by a chain or wire over a pulley to a counter-balance weight and pointer which is conveniently positioned against a calibrated scale.

9.2.2 Float Actuated Instruments

A development of the balanced float described above is the chain and float gauge consisting of a hollow float resting freely on the liquid surface and connected by a cord, chain or thin metallic tape over a pulley to a counterbalance weight. The float maintains a constant depth of immersion in a given liquid and rises and falls with any change in the liquid level. In so doing, it drives a pulley which operates an indicating, recording or control mechanism to show the changes in level (fig.9.3)

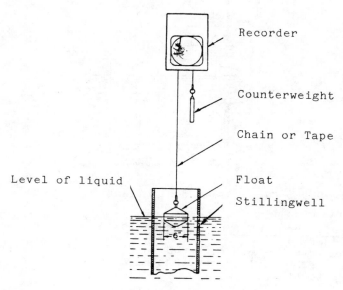

fig.9.3   Chain and float recorder

Turbulence in the liquid can be prevented from affecting the float by the addition of a stilling well around it. The pulley, over which the chain runs, actuates the instrument drive mechanism which, through gears and linkages operates a recording pen or indicator pointer against a chart or calibrated scale. The chain and float gauge is used on installations where the instrument can be mounted directly over the point of measurement. It is frequently used as a river gauge to record flow by measuring the head over a weir or flume. This device cannot be used for applications where the liquid is under pressure. Here some method of transferring the movement of the float through the container wall would be needed.

A caged float controller is used for pressurised applications (fig.9.4). A float and lever contained in a metallic cage, which is connected to the pressurised vessel, follows any variation in level. This movement is transmitted through the cage by a shaft rotating in a gland or stuffing box to a counterbalance lever outside the cage. This outside lever operates a pneumatic controller, or electrical switches or can be directly linked to a control valve regulating the flow of liquid into or out of the vessel.

140 Liquid level measurement

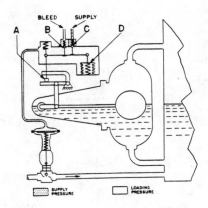

fig.9.4   Caged float controller

9.2.3 Displacement Type Instruments

The level detector, here, is a displacer, usually produced from a cylinder with closed ends which is pressure tight (fig.9.5).

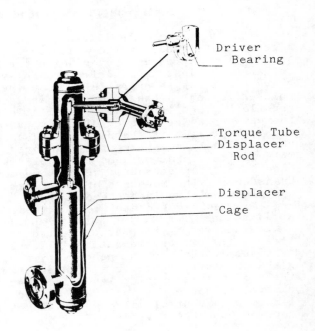

fig.9.5   Displacement type instrument

The displacer is denser than, and therefore sinks in the liquid being measured. The actual measurement made is the apparent weight of the displacer which reduces as the liquid level rises. The loss in weight is equal to the weight of liquid displaced, which in turn is governed by the volume of the displacer and the height of the level relative to the bottom of the displacer.

The weight of the displacer is measured by a torsion spring known as a torque tube assembly which transforms the weight variation into an angular movement of a torque tube shaft (fig.9.6).

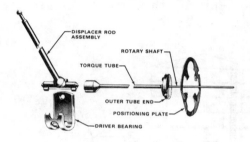

fig.9.6   Torque tube assembly

This angular movement can be used to drive a pneumatic or electronic transmitter or controller, producing an indicating or controlling output signal in direct proportion to the liquid level from the bottom of the displacer. Obviously, if the level rises above the top of the displacer no further change in weight takes place and therefore no further indication of level change is possible. The total variation in level measurement is therefore, governed by the height of the displacer.

Displacement units can also be used to measure the position of interface between two immiscible liquids having different specific gravities. This is commonly used to measure the interface between oil and water in a separator chamber to allow the oil and water to be drawn off the vessel individually. In this application it is essential that the displacer is always submerged in liquid.

Materials used for the construction, particularly the

displacer and torque tube assembly are carefully selected to combat the corrosive effect of the liquid or liquids being measured and the pressure and temperature of them, because this method of measurement like the caged float controller can be used for pressurised containers.

9.2.4 Hydrostatic Pressure Instruments

These instruments measure level by measuring the pressure exerted by the liquid on the measuring element. Various measuring elements can be used such as a pressure gauge, a bubble pipe, a pressure bulb or a force balance pressure transducer.

A pressure gauge directly connected to the discharge line from a storage tank can be calibrated to read directly the contents of the tank or liquid head above the gauge (fig.9.7).

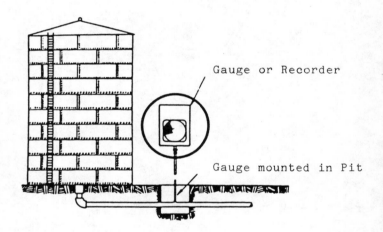

fig.9.7   Pressure gauge to record level

A bubble pipe level gauge can be used for many types of liquid regardless of its corrosive nature providing a suitable material can be chosen for the pipe. It is also suited for measuring the level of liquids carrying solids in suspension. In this instrument a small quantity of air or other non corrosive gas is allowed to bleed into a pipe lowered into the liquid (fig.9.8).

Liquid level measurement   143

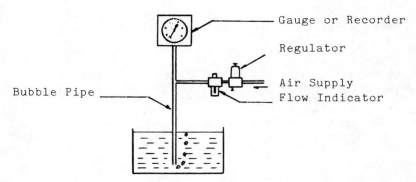

fig.9.8   Bubble pipe level gauge

The gauge measures the pressure of air needed to displace
the liquid in the pipe which is directly proportional to
the head of liquid above the lower end of the pipe. It
is imperative to always have a flow of air through the
bubble pipe, so there is usually a flow indicator built
into the system, and the pressure of air or gas needs to
be controlled at a value slightly higher than that needed
to balance the maximum level head intended to be measured.
The only disadvantage of this system is that as the level
falls the flow of air increases resulting in a high
consumption of compressed air.

A force balance pressure transducer also utilises air
pressure. In this type of gauge, level is measured by
means of a diaphragm which is exposed to the liquid
pressure on one side and pneumatic pressure on the other
side which is controlled to exactly balance the liquid
pressure (fig.9.9.).

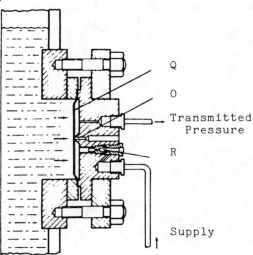

fig.9.9   Force balance diaphragm gauge

A pneumatic pressure greater than the liquid head to be measured is applied through the supply connection via an orifice 'R' to the right hand side of diaphragm 'Q' to oppose the force exerted by the liquid head. A bleed orifice 'O' is provided in the housing to vent the diaphragm chamber at a rate such that when the diaphragm is in equilibrium, the rate of air flow into the chamber equals the rate of flow out of it. Variations of pressure in the diaphragm chamber are measured by means of a pressure gauge or pressure receiver calibrated in terms of liquid level.

9.2.5 Differential Pressure Instruments

The hydrostatic pressure instruments previously referred to can only be used when the vessel to which they are applied is open to atmosphere. Where the container is pressurised a differential pressure measurement must be made, so that the container pressure can be subtracted from the total pressure beneath the liquid surface to obtain the actual level measurement. A simple mercury 'U' tube, manometer, can be used for this purpose (fig.9.10).

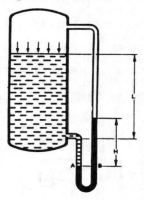

fig.9.10   A mercury 'U' tube used with a pressurised container

Here, the pressure of gas above the liquid is applied effectively to both legs of the mercury tube, so the height H of the mercury is directly proportioned to the height L of the liquid. This method enables the level to be measured at the bottom of the vessel which may be more accessible than a sight glass which would have to be viewed in line with the actual level surface.

9.2.6 Probe Methods.   The capacitance probe (fig.9.11)

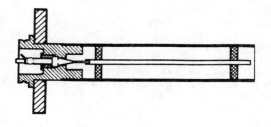

fig.9.11   Capacitance probe

is an electrostatic instrument measuring the change in capacitance of the probe when immersed in a liquid. The capacitance change can be measured by an electronic circuit adjusted to give level indication over the desired range. The probe usually consists of two concentric tubes where the capacitance is a function of length and diameter of the tubes and the dielectric constant of the material between the tubes. The variation in capacitance is measured and converted into direct current readings. This method is suitable for most liquids other than those which would separate out on standing into a conductor and a non conductor. It is also unsuitable for conducting liquids which froth excessively or where solids contained in the liquid could deposit out.

Another type of probe instrument used only to indicate one set position of level is the vibrating probe (fig.9.12).

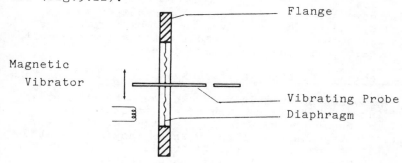

fig.9.12   Vibrating probe

The probe unit consists of a detecting device, a control amplifier and the necessary interconnecting wiring and mains supply. The probe is mounted horizontally at the required level on the tank or hopper wall such that the fluid or granular material in the tank can come in contact with the probe.

The probe assembly consists of metal rod passing through and fixed in the centre of a thin metal diaphragm welded into the bore of a flange, allowing both ends of the rod free to vibrate. One end of the probe projects into the tank while the other end is housed outside the tank in conjunction with a magnet coil assembly used to vibrate the probe.

While the tank contents is below the probe it is free to vibrate but when the level reaches the probe it is prevented from vibrating and stops the driving circuit from oscillating. This, through a control relay, and amplifier circuit can initiate a control device to empty or fill the tank or it can operate an indicator lamp or alarm device.

9.2.7 Radio Frequency Method

This method of level measurement operates by measuring the position of a small sensing element emitting radio frequency signals from an antenna which is maintained by a servo mechanism about 2mm above the liquid surface.

The probe is linked to the servo mechanism by a perforated stainless steel tape and the movement of the servo mechanism provides an indication of liquid level.

This method, although expensive is very accurate and is unaffected by liquid specific gravity and has applications on large storage tanks handling corrosive liquids either at atmospheric or pressurised condition.

9.3. SUMMARY

9.3.1. Visual Indicators

Visual methods of level measurement are used where no direct control of the process is required and where periodic readings of the level can be made to ensure it is within certain limits. Visual methods are the most economical of all the measuring devices.

9.3.2. Float Actuated Instruments

Various types within this range can be chosen depending on whether the liquid being measured is operating under pressure or is open to the atmosphere. The latter types are limited to where easy access above the point of measurement exists and is typically used as a river gauge. It has the advantage that it can be linked to a recorder to enable permanent records of changes to be made. The pressurised units have a certain limitation of span but have the advantage of the facility of direct control of the process if required.

9.3.3. Displacement Type Instruments

This is a versatile method of liquid level measurement which may be used for pressurised containers. It produces either a pneumatic or electric control signal which may be used to operate a control valve or an indicator or recorder located remote from the point of measurement.

9.3.4. Hydrostatic Pressure Instruments

Various methods in this category, ranging from a simple pressure gauge to a sophisticated force balance transducer, cover a wide range of applications. Most of these types can be used where the liquid is corrosive, providing a suitable material for the sensing element is chosen.

9.3.5. Differential Pressure Instruments

The same range of devices and applications apply to this category as to the previous one. In most instances it is accomplished by using two devices, one at the top and one at the bottom of the vessel.

9.3.6. Probe Methods

The capacitance probe referred to is limited to liquid applications where the liquid remains stable and is of a conducting type. The vibrating probe is only used to indicate a particular level attainment but can also be used for granular materials.

9.3.7. Radio Frequency Methods

A rather expensive but very accurate method which is unaffected by the liquid specific gravity or corrosive nature, since the sensing element never comes into contact with the liquid.

Chapter 10

Control valves and actuators

B. Baldwin

10.1 INTRODUCTION

A control valve is a device which regulates the flow of a fluid in a pipe line. It consists of two elements, a restricting element and an actuating element. The actuating element commonly known as an actuator, transforms a control signal from a controller into a motivation of the restricting element which, in turn, regulates the flow of fluid in the pipe line. The control valve is often referred to as the last link in the automatic control system, and is probably the most important individual unit in the loop. All the effort of measuring the variable, transmitting and comparing its value with the desired value, and initiating a control signal would be wasted if the control valve did not make a correction to the process, in accordance with the response corrections called for by the controller.

Nobody really knows when the first control valve was produced, but there is evidence that the Romans used valves for the control of water flow. These were usually made from wood and were manually operated, so the actuators were in that case humans. The advent of the Industrial Revolution brought with it many instances where control valves were needed to be automatically controlled. Instead of having a hit-and-miss method of someone watching a pressure gauge or level measurement, and turning a handwheel to close a valve down or open it up, there needed to be an immediate link between the measurement taken and the operation of the valve to reduce time lapse, and avoid unnecessary waste in both time and valuable process fluid.

During the early 1900s, sales of petrol driven motor cars and then the advent of World War I, heralded a growth in the petroleum industry, where more control valves were needed to maintain the continuous processes being developed. Gradually during the 1920 to 1930 era, new process industries which required more sophisticated process control equipment were developed, and the basis of the present day range of control valves and associated control instrumentation was born.

10.2 THE RESTRICTING ELEMENT

This part of the control valve which is responsible for regulating the fluid flow is available in many forms. The range currently used in the control industry falls basically into two categories; sliding stem and rotary.

10.2.1 Sliding Stem Valves

The valve has to be capable of controlling the flow of fluid passing through a pipeline. It must therefore have a housing known as the valve body, which can be conveniently fitted into the pipeline. To achieve this the valve body is either flanged, screwed or has welded connections.

To vary the flow of fluid through the valve, there needs to be an adjustable passage within the valve. This is provided by a seat ring and a moveable valve plug (fig.10.1).

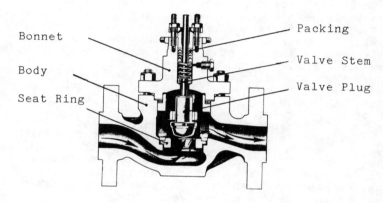

Fig.10.1 Flow passage through a valve

The valve plug is connected via a stem to the actuator and therefore moves in accordance with the dictates of the control signal. The shape of the valve plug is very important, because this determines the amount of flow change which will develop in response to a specific linear movement of the actuator. This relationship between the flow through the valve and the linear movement produced from the control signal, is referred to as the valve characteristic.

10.2.1.1 Characteristics.
There are three basic characteristics used; linear, equal percentage and quick opening. (fig.10.2)

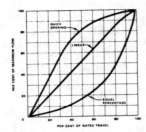

Fig.10.2   Inherent flow characteristic curves

The linear characteristic produces a proportional increase or decrease in flow for each incremental change in the control signal, whereas the equal percentage characteristic produces a percentage increase or decrease in flow, which means that a change in control signal when the plug is close to the seat, will produce a much smaller change in flow than would be produced by the same change in signal at a point where the valve plug is some way off from the valve seat. The quick opening characteristic is opposite to that of the equal percentage, producing a large change in flow for a small lift of the plug from the seat.

The selection of the ideal characteristic for a control valve depends upon the application for which the valve is being used. The installed characteristic, which is obtained when the valve is in actual use, may well be different to the inherent characteristic, since the pressure drop across the valve producing flow may well change as the opening of the valve changes.

The linear characteristic is usually selected for liquid level control and for flow control where a constant gain is required. The equal percentage characteristic is used on pressure control applications and where a large percentage of the pressure drop is absorbed by the system as a whole with only a small percentage available at the control valve. The quick opening characteristic is often used for relief valve applications where the valve needs to open rapidly and pass a high flow with a small lift.

10.2.1.2 Bonnets. The actuator movement is transmitted to the valve plug through the valve stem, which must pass through the pressure envelope of the body assembly. This is achieved by the bonnet which allows unrestricted linear movement of the valve stem, through an area known as the gland or packing box. Here, packing in the form of moulded or square section rings, prevents loss of fluid from the bonnet, and allows the stem to move with little frictional resistance. For hazardous or very expensive fluids where any leakage which may occur through the packing would be undesirable, a special seal called a bellows seal can be used to replace the packing.

10.2.1.3 Size range. Sizes of valves in this sliding stem category, range from 12mm up to 400mm. This size denotes the size of the pipeline connection. They can be fitted with valve trims, that is, the seat and plug size, of equal size to the body, or of a smaller size for reduced flow requirements, where the inlet velocity to the valve is required to be kept low.

10.2.1.4 Body forms. The valve body can take various forms. It can be single or double ported, straight through or angled, or three-way.

A single ported valve has the appearance of a domestic water tap, having one variable passage to the flow stream. (fig.10.1) This, especially in the larger sizes, has limited applications involving only low pressures, because high pressures would need correspondingly higher actuator pressures to force the valve plug closed. High pressure applications are generally accommodated by a balanced construction of valve, such as a double ported construction where the flow stream is passed up through one opening and down through the other, producing a balance of forces. (fig.10.3) The flow direction through most valves is normally kept in a 'flow to open' direction. This is especially important with single ported valves, in order to minimise the slamming effect and consequent damage, which would occur if flow to close operation were used.

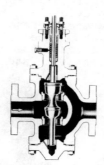

Fig.10.3   A double ported valve

Three-way valves are used where two flow streams need to be mixed in variable quantities, or where one stream needs to be split into two pipelines, again with the facility to vary the amount selected for each line. (fig.10.4). This type of valve is often used on temperature control applications.

152 Control valves and actuators

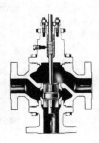 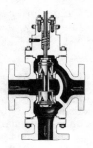

Fig.10.4   Three-way valves used for converging or diverging flows

10.2.2 Rotary Valves

In rotary valves the restricting element moves in a rotary path relative to the valve body. There are various forms, known as ball valves, butterfly valves and eccentric disc valves.

Every valve, due to the formation of turbulence, noise, heat etc. produces a loss in pressure in the pipeline known as a pressure drop. Rotary valves in general cause less restriction than sliding stem valves and are referred to as high recovery valves.

10.2.2.1 Ball valves. These have a rotating, hollowed out ball which produces a varying flow passage, depending on its position. (fig.10.5). When in the wide open position, a ball valve provides negligible resistance to the flow path and therefore creates a very small pressure drop. The action of the ball when closing has a chopping effect which is very useful when handling fluids such as slurries or paper stock.

Fig.10.5   A V-notch ball valve

10.2.2.2 Butterfly valves.
These have a rotating disc which rotates within a wafer body, such that when the disc is in the open position it projects into the pipeline (fig.10.6).

Fig.10.6   A typical butterfly valve

One must ensure therefore that the valve is in the closed position before attempting to remove this type of valve from the line. The disc of the butterfly valve rotates about its centre and when closed it is difficult to achieve a shut-off condition with a very low leakage path. This problem lead to the development of the eccentric disc valve, which has a centre of rotation off-set from the centre of the disc, allowing it to close tight against its seal ring and also having the advantage of reducing wear between the disc and the body of the valve. (fig.10.7)

Fig.10.7   An eccentric disc valve

154 Control valves and actuators

10.2.2.3 Size range. The sizes of rotary valves available range from 25mm (1 inch) to 600mm (24 inch) for ball valves, and from 50mm (2 inch) to 900mm (36 inch) for butterfly valves.

10.2.2.4 Characteristics. The characteristic of rotary valves cannot be varied as they can for sliding stem valves, and approach an equal percentage characteristic for butterfly and eccentric disc valves and approximately linear characteristic for ball valves.

10.2.3 Materials

The materials chosen for the valve body in both sliding stem and rotary valves has to be strong enough to withstand the pressures and temperatures exerted on the valve in both test and operating conditions, and must resist the corrosion and erosion effects of the flowing fluid. A large majority of control valve bodies are made from high-tensile cast iron or cast carbon steel. But alloy steels, stainless steels and special materials need to be selected for handling corrosive fluids, or where the flowing temperature is abnormally low or high. The valve plug, seat rings, valve stem, guide bushings and packing parts, which are often referred to as the valve trim, are usually manufactured in stainless steel, but again special materials need to be selcted for very corrosive applications. Hard surfacing materials such as cobalt alloys are often applied to valve seats and guides, and are an economical solution for services handling high temperatures or pressure drops.

10.2.4 Valve sizing

It can be shown from Bernouli's energy equation that the relation between the flow and the pressure drop across a restriction in a pipe follows a square root law. We can therefore state that the flow is proportional to a constant multiplied by the square root of the pressure drop across the valve. This constant, the valve technologists use to enable the sizing of control valves, and, for liquid flow, is known as the Cv of the valve. The equation used for liquid sizing is:

$$C_v = Q\sqrt{\frac{G}{\Delta P}}$$

Equations for gas and steam sizing are similar in construction but involve more factors and result in a Cg value for gases and a Cs value for steam. Practical tests carried out on each type and size of valve determine a usable sizing co-efficient for each valve which is listed by manufacturers for comparison.

10.3 VALVE ACTUATORS

The valve actuator is that part of the control valve which accepts a signal from the controller, and uses this signal to position the restricting element of the valve to control the fluid passing through the valve body.

Different types of valve actuators are used for operating control valves. The commonest type in use is the pneumatic spring opposed diaphragm actuator. Other types are pneumatic piston actuators and electric actuators.

A valve actuator has to operate satisfactorily in conjunction with a control valve operating in a control loop under all service requirements. There should normally be a linear relationship between the control signal and the output movement of the actuator, hysteresis of movement must be negligible or kept to an absolute minimum, and the actuator must be sufficiently stiff to withstand the operating forces which arise from operating the control valve. The design must also be rugged enough to withstand the stresses encountered during shipment as well as during normal plant operation.

10.3.1 Pneumatic Diaphragm Actuators

Because of its simplicity, the pneumatic spring opposed diaphragm actuator is by far the most widely used type. It is used in conjunction with both sliding stem and rotary valves, and finds applications on all types of valve bodies, except those in which the unbalanced forces on the valve are so great that the power requirements of the spring opposed diaphragm actuator make it unwieldy or impracticable.

A typical pneumatic spring opposed diaphragm actuator is shown below (fig.10.8). It consists of a moulded rubber diaphragm contained within diaphragm casings, and opposed by a loading spring mounted on a connecting yoke. The movement of the diaphragm is transmitted through a diaphragm plate driving an actuator stem connected to the control valve stem.

Fig.10.8 A typical pneumatic diaphragm actuator

The control signal applied to a diaphragm actuator is in the form of an air pressure, usually in the range of 0.2 to 1 bar such that the actuator stem starts to move at 0.2 bar and completes its travel at a pressure of 1 bar.

Diaphragm actuators are available in various sizes which are selected to suit the size of the valve and the operating pressure conditions within the valve.

The actuator action can be selected such that increasing air pressure to the actuator pushes the actuator stem down, which, on a push down to close valve, would result in increasing air pressure closing the valve, or it can have the reverse action, such that increasing air pressure opens the valve. (fig.10.9). The type of actuator chosen usually depends on whether the valve should open or close in the event or air signal failure, which can be very important when considering the safety of the overall system.

Fig.10.9   A reverse acting diaphragm actuator

Manual operators or handwheels can be fitted to diaphragm actuators to provide a means of manually positioning the valve plug during an emergency, during start-up or in the event of air failure. Two types of handwheel are generally used, either a handwheel mounted on the top of the valve actuator, or a side mounted handwheel mounted on the yoke of the valve actuator.

10.3.1.1 Valve positioners.   Although the diaphragm actuated control valve is generally designed with sufficient force to position the valve accurately in proportion to the change in instrument signal, under difficult service conditions sufficient force may not be available. Where such conditions exist, a valve positioner or booster relay should be used in conjunction with the control valve to ensure accurate and dependable positioning of the valve. (fig.10.10).

Fig.10.10 A valve positioner mounted on a reverse acting actuator

The valve positioner is a pneumatic positioning device mounted on the valve actuator yoke with a mechanical connection to the valve stem, and operates to modulate the air pressure on the valve actuator diaphragm until the valve stem is positioned in accordance with the demands of the control signal. Reference to a schematic of a typical valve positioner will clarify the operation (fig.10.11).

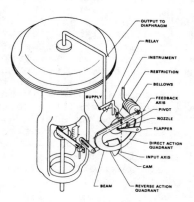

Fig.10.11 Schematic illustration of a valve positioner

Air pressure is supplied to the relay supply point and a fixed restriction. A flapper moves against a nozzle which is connected to the fixed restriction. The diameter of the fixed restriction is less than the diameter of the nozzle so that air can bleed out faster than it is being supplied when the flapper is not restricting the nozzle. When the instrument pressure increases, the bellows expands to move the beam, causing the flapper to restrict the nozzle. The nozzle pressure increases and moves a relay diaphragm assembly to open a supply valve within the relay. This allows the output pressure to the diaphragm casing of the control valve to increase, moving the actuator stem downward. Stem movement is fed back to the beam by means of the cam which causes the flapper to move away from the nozzle. Nozzle pressure decreases and the relay supply valve closes to prevent any further increase in output pressure. The positioner is once again in equilibrium but at a higher instrument pressure and a new valve plug position.

When the control instrument pressure decreases, the bellows contracts, aided by an internal range spring to move the beam and uncover the nozzle. Through relay operation an exhaust valve in the relay opens to release the diaphragm pressure to atmosphere, permitting the actuator stem to move upward. Stem movement is fed back to the beam by the cam to reposition the beam and flapper. When equilibrium conditions are obtained the exhaust valve closes to prevent any further decrease in diaphragm casing pressure. For each value of control signal, therefore, there is a finite value of valve stem position which will always be established, regardless of external force variations applied to the valve stem.

10.3.2 Piston Actuators

When the application is such that the valve actuator is required to operate against heavy out of balance forces caused by unbalanced pressures, or the weight of the valve, or is handling high viscosity fluids, the valve actuator is required to develop a thrust greater than that which can be conveniently supplied by a diaphragm actuator. In these circumstances, a piston actuator may be used to operate the control valve.

The piston actuator is generally pneumatically operated and is either integrally mounted on the valve (fig.10.12) or is furnished as a separately mounted power cylinder for use with large butterfly valves. The actuator consists of a double acting piston in a cylinder operating on an air supply of up to 10 bar, fed to it through a valve positioner, working on an instrument signal of 0.2 to 1 bar.

Fig.10.12   A piston actuator

The piston actuator relies on a difference in air pressure on either side of a piston to cause movement of its actuating stem to stroke the control valve. As the piston is not spring loaded and operates at a much higher pressure than that furnished by the controlling signal, it is necessary to incorporate a valve positioner in the operating mechanism to act as a relay and feedback device to control the movement of the piston and hence the valve opening.   To understand the operation of the piston actuator refer to the schematic (fig.10.13)

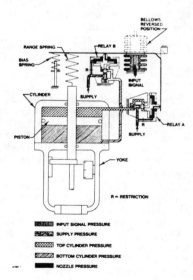

Fig.10.13   Schematic illustration of a piston actuator

The pneumatic signal from a controller is fed to the bellows of the positioner. On an increase in signal pressure the bellows expands and moves a beam which pivots around a fixed point simultaneously to uncover the nozzle of the air relay "B" and cover that of relay "A". The nozzle pressure of relay "A" increases and through relay action causes the cylinder pressure over the top of the piston to increase. At the same time, the nozzle pressure in relay "B" decreases, causing the cylinder pressure below the piston to similarly decrease.

The unbalanced pressures acting on the piston cause it to move downwards to change the valve plug position. The movement of the piston is fed back to the beam by means of a range spring connected between the beam and an extension of the piston rod. This arrangement provides feed-back to the system to prevent over-correction and ensures a definite position for the piston for every value of instrument signal.

Piston actuators can be arranged to either move the valve stem up or down on increasing instrument signal, depending on the position of the instrument bellows.

In order to move the piston to a desired safe position in the event of air supply failure, piston actuators may be fitted with a spring to return the actuator stem to the up or down position. This has a disadvantage however of absorbing a certain amount of available thrust in compressing the spring. It is also possible to fit pneumatic tripping devices with air pressure stored in capacity chambers, which can be switched to either move the piston to the up or down position in the event of air failure, or to fit a device which locks the piston in the last controlled position.

As with diaphragm actuators, it is possible to fit handwheels to piston actuators to manually position the valve during start-up or emergency shutdown conditions.

10.3.3 Electric Actuators

Electric actuators usually consist of an electric motor with gear trains arranged to provide a wide range of output forces. They are often used in remote locations where no other power source is available. The majority of electric actuators are used for valves requiring only on/off service, but can be used for modulating control with the addition of a potentiometric feed-back device. Relative to diaphragm and piston actuators they are quite slow in operation. The only fail-safe action available is a lock in last position, or fail fix, one. They can be fitted with limit switches and torque limiting devices and have long stroke capabilities, which makes them suitable for very large butterfly valves.

Closely connected to the elctric actuators are the electro-hydraulic type (fig.10.14) which again only need an electricity supply but which will produce excellent throttling control in response to a milliamp signal source.

Fig.10.14   An electro-hydraulic actuator

An electric motor is connected to a pump feeding oil at high pressure to a piston, which is positioned in a similar manner to the pneumatic piston in response to the control signal.

10.3.4   Actuator Selection

The choice of actuator for a specific application will be dependent upon various factors.

1. The power source available.
2. Fail safe requirements.
3. Force requirements of the valve.
4. Form of control required.
5. Cost.

10.3.4.1 Power source. Most plants have both compressed air and electricity available on site, but there are occasions where the site is in a remote location, and it is uneconomic to install an air compressor. This will automatically cut down the choice to either electric or electro-hydraulic.

10.3.4.2 Fail safe requirements. Although both pneumatic and electrical systems are very reliable, it is always necessary to consider the effect of the loss of the power source, which could result in a hazardous situation developing. By storing energy in springs or in pneumatic capacity chambers, this reserve power can be called upon to move the valve to its safe position when the main power source fails. The three failure modes available are: fail-open, fail close, or fail-fixed which locks the valve in its last controlled position.

10.3.4.3 Force requirements of the valve. An actuator must have sufficient output thrust to handle the requirements of the valve, although it is obviously uneconomic to fit an actuator which is over-powered. It has to be capable of closing the valve under all possible service conditions. There is usually a wide choice of sizes available for all types of actuators from pneumatic spring and diaphragm through to electro-hydraulic, and the actual selection of size is usually best left to the manufacturer who will choose the one most suited to the particular valve.

10.3.4.4 Form of control required. The form of control can be either two position (on/off) or analogue (throttling). On/off actuators receive signals of either zero or maximum from the controlling instrument, and so the choice of actuator is fairly simple. It merely has to possess sufficient force to close and open the valve. Throttling actuators require much more careful selection. Here the actuator has to respond to every change in signal from the controller, which is continuously monitoring the process. To achieve this the actuator must either be directly compatible with the controlling signal, or some intermediate device such as a relay or positioner must be added.

The speed of action and the effects of vibration and temperature must also be considered, although stroking speed is not normally a problem, it may be necessary to close or open a valve within one or two seconds in emergency conditions. When considering pneumatic actuators, the piston actuator will travel at much faster speeds than the spring and diaphragm type unless special booster relays are installed.

10.3.4.5 Cost. The cost of an actuator is very often the over-riding factor in its selection. Where compressed air is available, the most economical, both in initial cost and subsequent maintenance, is the diaphragm actuator. Where the thrust capability of this type is insufficient, the next best choice is the piston actuator.

Electric or electro-hydraulic actuators should only be considered when compressed air is not available or when a very high force is required, which eliminates the pneumatic type.

Chapter 11

Errors and uncertainty in measurements and instruments

Professor L. Finkelstein

11.1 INTRODUCTION

An error in a measurement is the difference between the result of a measurement and the true value of the measured quantity. It arises in all measurements due to the fact that procedures and instruments of measurement are not in practice perfect. The error in a measurement is not in practice precisely determinable so that the result of a measurement is accompanied by a degree of uncertainty.

The article will consider the nature, sources and methods of description of errors and uncertainty in measurements and in measuring instruments.

11.2 DEFINITION OF ERROR

Error ε is the difference between the result of a measurement q_r and the true value of the measured quantity q_m:

$$\varepsilon = q_r - q_m \qquad (11.1)$$

The true value of a quantity, that is the number which corresponds to the quantity under perfect measurement, as defined by the appropriate measurement scale, is an idealised concept which cannot be empirically determined. Even the definitions of the SI scales involve uncertainty. In general therefore, when we speak of true value we mean the conventional true value, which is a value approximating to the true value, such that for the purpose for which that value is used, the difference between the two values can be neglected. For example in the calibration of an industrial instrument of moderate accuracy the measured value of a quantity by an instrument of high accuracy is accepted as the conventional true value.

11.3 MEASURED SYSTEM AND MEASURING SYSTEM ERRORS

Consider a measurement system as shown in Figure 11.1.
The total measurement system consists of the object or system under measurement and the measuring instrument system, which includes both the instrument acquiring the measurement

information and the system processing it.

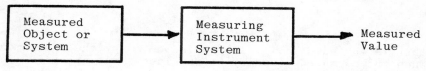

Fig.11.1   Measuring System.

The total measurement error consists of two components:
(i) the error originating in the measured system and the interface with the measuring instrument systems.
(ii) the error originating in the measuring instrument system, also known as instrumental error.

11.4 ERRORS ARISING IN THE MEASURED SYSTEM

Consider the errors originating in the measured system and the interface. Let us consider as an example the measurement of the exhaust gas temperature of a small gas engine at some nominal speed and load.

The first type of error arises due to the imperfection of the model of the measured system which underlies the measurement process. The temperature of the exhaust gases may be uniform, and a result of measurement which assumes uniform distribution is in error.

The second type of error arises due to the change in the configuration of the measured system produced by the introduction of the sensor. The introduction of a sensor in the exhaust duct raises the back pressure of the engine, requiring an increase in fuel flow and hence an increase in exhaust gas temperature.

The third type of error is due to the power absorption from the measured system by the sensor. In this case the heat exchange between the exhaust gases and the sensor alters the temperature of the exhaust gases.

The above discussion has only dealt with the errors in passive measurement, in which the measured variable is sensed. The errors arising in a measured object in the process of active measurement will be considered later.

11.5 DEFINITION AND CLASSIFICATION OF MEASURING SYSTEM OR INSTRUMENTAL ERRORS

An instrument functions by maintaining a functional relation between the information carrying characteristics of the signal or signals at its input and the information carrying characteristics of its output. In a measuring instrument a functional relation is maintained between the magnitude of the measured quantity at the input and indicated numbers at the output. There is a nominal or intended instrument response law which relates the instrument output numbers to the value of the measured quantity at the input.

The nominal response law is:

$$q_i = q_m \quad (11.2)$$

that is the numerical value q_i indicated by the instrument when the measured quantity of the instrument q_m is applied at the input, is equal to the true value of q_m.

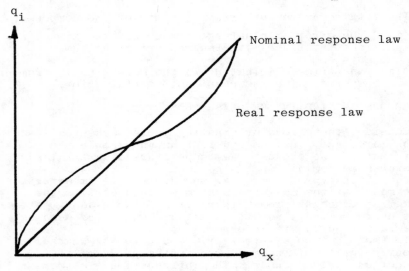

Fig.11.2   Response Laws.

The nominal response law is determined by calibration or by mathematical modelling and is deemed to hold for specified conditions of the environment and the system of which the instrument is part, termed reference conditions.

In practice the real response law of an instrument is such that the application to the instrument at the input of a true value q_m results in an indicated value q_i, which is not equal to q_m, so that we have:

$$\varepsilon_i = q_i - q_m \quad (11.3)$$

ε_i is the instrumental error.

There are two components of instrumental errors:

    (i)   <u>Intrinsic errors</u>: these are errors of the instrument when used under reference conditions.

    (ii)   <u>Influence errors</u>: these are errors which arise from the action on the instrument of influence variables, that is physical variables in the instrument environment, which are not under measurement but which affect the instrument response. They are due to the departure of the influence variables from reference conditions.

11.5 INTRINSIC INSTRUMENTAL ERRORS

There are two types of real instrument response under reference conditions:
- (i) determinate - that is one in which repeated applications of a particular input always result in the same response.
- (ii) indeterminate - that is one in which repeated applications of a particular measured quantity input result in different responses.

If the real response is determinate but differs from the intended one it is termed false and results in a determinate error. We may distinguish the following sources of error in a false response law:
- (i) Errors in the determination of the intended response law arising either from:
  - calibration, when erroneous or uncertain standards or procedures are used to establish the intended response law.
  - calculation, when the intended response law is calculated using a mathematical model which is inadequate for the purpose.
- (ii) Errors resulting from accepting a nominal response law different from a determined or determinable real one. Thus, for example, a mass produced instrument may be fitted with a linear scale correct at two points only, though it is known that the real instrument response is non-linear.
- (iii) Errors resulting from a change of the real response law of the instrument with time as a result of ageing of material properties, wear, or damage due to maloperation. They are known as secular.

The following are some of the most common forms of false response law errors.
- (i) Zero error - the deviation from zero of the indication of a measuring instrument for zero value of the measured quantity.
- (ii) Gain error - in a linear real response law, the deviation of the slope from the unit value assumed in the nominal law.
- (iii) Non-linearity error - the deviation of the real response from nominal linear response. Various methods are employed for the expression of non-linearity.
  - departure from independent linearity, that is deviation from a best fit straight line by least squares for the available data.
  - departure from zero-based linearity - that is deviation from a best fit straight line for the available data but assuming no zero error.

168   Errors and uncertainty

> – departure from terminal-based linearity, that is deviation from a straight line joining the responses of the instrument at the terminal points of the span.

The are shown in Figure 11.3

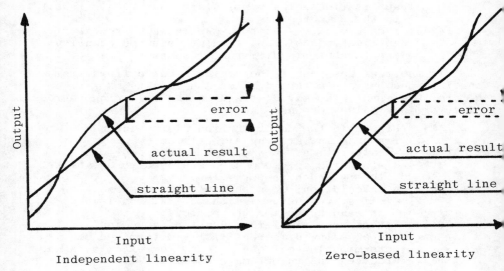

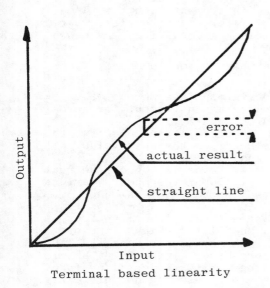

Fig.11.3   Linearity.

(iv) Hysteresis error – the error resulting form hysteresis, that is the property of a measuring instrument whereby it gives different indications according to whether that value has been reached by a continuously increasing or continuously decreasing change of that quantity.
(v) Quantisation error – the error which may result from the measurement of a value of a quantity by a process in which response can only change in discrete quantum steps such as in digital measurement.
(vi) Dynamic error – the error in the operation of an instrument used in the dynamic mode in which the instantaneous value of the instrument indication is required to be a function of the instantaneous value of the measured quantity, where that quantity is varying. Practical instruments always have a memory or lags as a result of which their instantaneous indication is a function not only of the instantaneous measured quantity input but depends also on the past history of the measured quantity. Typically a step change in the measured quantity results in either a slow rise in instrument indication to a new steady state value corresponding to the changed measured quantity, or an indication which oscillates about the steady value before settling down to it. When the measured quantity varies harmonically the indication of a practical instrument tends to lag behind it in phase and the gain of the instrument varies with frequency. The difference between the instantaneous indication and the instantaneous measured value is the dynamic error.

Indeterminate response laws originate in such effects as solid friction, varying resistance at contacts and the like, which result in a non-repeatable relation between measured quantity and instrument indication. Indeterminate response laws give rise to random errors which will be considered later in the article.

11.7. INSTRUMENTAL INFLUENCE ERRORS

Instrumental influence errors may be caused by influences arising in:
(i) the measuring system environment.
or else in the measuring instrument super-system that is:
(ii) the instrument output interface – the influence being the effect on the instrument indication of any power abstracted from the instrument at its output.
(iii) the instrument input interface – the influence being the effect on the measured system of the power abstracted from the measured system by the instrument. The resultant error is strictly a

measured system error rather than an instrumental one.
(iv) the energising power supply in active instruments - the power supply being an influence on instrument indication, and its departure from reference conditions causing error.

The environment influence variables which can act on an instrument may be classified according to their energy form and the effects they have reviewed. The variables may be:

(i) Mechanical. Vibrations affect the relative position of elements of an instrument, introduce variable stresses which affect material properties and so on. The force of gravity acts on components and hence if the orientation of a mechanical instrument varies with respect to the vertical the response may be affected.

(ii) Electrical. Electrical signals are induced in circuits by various forms of stray coupling.

(iii) Fluid mechanical. The effect of high ambient pressures may change the geometry and material properties of an instrument and hence the instrument response law. This may be of significance in, say, underwater instrumentation.

(iv) Thermal. Temperature affects all dimensions and material properties. Hence temperature is a significant influence variable for all instruments.

(v) Chemical. The change in chemical composition of any instrument elements as a result of environmental chemical influence may significantly affect their behaviour. Electrochemical potentials may also be significant.

(vi) Ionising radiation. The effect of such radiation may alter material properties of instrument components or induce electrical signals.

Let us denote the measured quantity by q_m, and the indicated value by q_i. Let us further denote an influence (disturbance) variable by $q_{d1}, q_{d2}, \ldots, q_{dn}$. Then the response law of the instrument may be written as:

$$q_i = f(q_m, q_{d1}, \ldots) \qquad (11.4)$$

Let the reference conditions be denoted by $\bar{q}_{d1}, \bar{q}_{d2}, \ldots$ Then, in general, the influence error ε_d is given by:

$$\varepsilon_d = f(q_m, q_{d1}, q_{d2}, \ldots) - f(q_m, \bar{q}_{d1}, \bar{q}_{d2}, \ldots) \qquad (11.5)$$

Then writing:

$$\Delta q_{dn} = q_{dn} - \bar{q}_{dn} \qquad (11.6)$$

We have:

$$\varepsilon_d = \left(\frac{\partial q_i}{\partial q_{d1}}\right) \Delta q_{d1} + \left(\frac{\partial q_i}{\partial q_{d2}}\right) \Delta q_{d2} + \ldots \quad (11.7)$$

where if the values Δq_{dn} are small the terms with higher order derivatives may be neglected. In many practical cases these derivates tend to be negligibly small. $\partial q_i/\partial q_{dn}$ are termed the influence error coefficients. It is more usual to specify the relative coefficients $(1/q_i)(\partial q_i/q_{dn})$. The coefficients q_i/q_{dn} may be independent of q_m, so that the influence errors are additive, say in the case of inductive pickup, resulting typically in a zero error with no change of gain. In other cases the coefficients are functions of q_m resulting in a change of gain as a function of the influence variables. A typical example would be the change of gain with temperature of a load cell employing an elastic deformation element.

The above considerations apply to reversible effects of influence variables. Irreversible effects are not influence but secular intrinsic errors.

11.8 TIME VARIATION OF INFLUENCE AND INTRINSIC ERRORS

The time variations of the effects of influence variables are of importance. We distinguish between:

(i) Short term variation - that is those in which the influence variables change appreciably over the period of a particular measurement process.
(ii) Long term variation - that is changes of influence variables so slow that the variables are effectively constant over a particular measurement process.

Similar considerations apply to secular changes of intrinsic errors.

11.9 SYSTEMATIC AND RANDOM ERRORS

We distinguish between:

(i) <u>Systematic erros</u>, that is errors which are the same in repeated measurements of the same value of a measured quantity.
(ii) <u>Random errors</u>, that is errors which differ in repeated measurements of the same value of a measured quantity.

Systematic errors are due to:
- (i) Intrinsic errors arising from a determinate false law.
- (ii) Influence errors arising from long term influence variations or secular intrinsic instrument changes, which although basically random, are effectively systematic for a particular measurement.

Random errors are due to:
- (i) Intrinsic errors arising from an indeterminate response law.
- (ii) Influence errors arising from short term variations of an influence variable.

With respect to systematic errors we may distinguish between:
- (i) Certain systematic errors.
- (ii) Uncertain systematic errors.

Certain systematic errors arise when we have a known real instrument law but where the nominal response law is different. Alternatively they may arise where we know the deviation of influence quantities from reference conditions, and the law relating the influence variables to the instrument response. Certain systematic variables can be compensated by correction.

Uncertain systematic errors arise when the real response law is determinate but unknown, for reasons explained above.

11.10 UNCERTAINTY AND ITS SPECIFICATION

Certain systematic error is, by definition, a known quantity at every value of the measured quantity.

Random errors and uncertain systematic errors are unknown quantities. They mean that no value of the measured quantity can be assigned with certainty to the result of a measurement.

Given a particular value q_i indicated by an instrument under reference conditions the true value of the measured quantity q_m is related to q_i by a conditional probability density function:

$$p(q_m \mid q_i) \qquad (11.8)$$

A measure of dispersion of this p.d.f. is a measure of the inherent instrumental uncertainty at an indicated value of q_i. The p.d.f. can in theory be estimated by calibration.

In practice it is assumed that the random errors for any measured value are distributed according to a Normal distribution with a zero mean error: thus with the notation used:

$$p(\varepsilon) = \frac{1}{\sigma\sqrt{2}} \exp\left(-\frac{\varepsilon^2}{2\sigma^2}\right) \qquad (11.9)$$

where

$$\varepsilon = q_i - q_m \qquad (11.10)$$

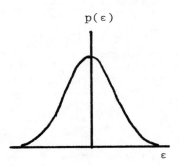

Fig. 11.4 Normal error distribution.

The reason for the assumption of zero mean error is than an intrinsic random error with a meandifferent from zero can be taken to be an uncertain systematic error.

The assumption of a Normal distribution is based on the supposition that in a well designed measurement random intrinsic errors will be the result of a large number of uncorrelated component errors each making a small contribution to the resultant error. Such a process will, according to the Central Limit Theorem result in a Normal error distribution. This assumption does not hold for many less precise measurements, for which the random error is the result of a small number of error components possibly correlated, each making a substantial contribution to the error. Nevertheless the assumption of a Normal error distribution is adequate for most practical purposes.

σ the standard deviation of the error distribution - is a measure of the uncertainty of the measurement given a Normal distribution. In practice σ is estimated experimentally by replicating the measurement of a particular value of the measured quantity.

Given a set of n indicated values $q_{i1}, q_{i2}, \ldots, q_{in}$ for a particular value of q_m the best estimate of q_m is:

174 Errors and uncertainty

$$\hat{q}_m = \bar{q}_i = \frac{1}{n} \sum_{k=1}^{n} q_{ik} \qquad (11.11)$$

The best estimate of σ^2 is given by:

$$\hat{\sigma}^2 = \frac{1}{n-1} \sum_{k=1}^{n} (q_{ik} - \bar{q}_i) \qquad (11.12)$$

Because the estimation of $\hat{q}_m$ by $\bar{q}_i$ was based on a limited number of observations, repeat determination of $\bar{q}_i$ would produce different values. For large values n these values will be distributed according to a Normal distribution. The best estimate of the variance of the latter distribution is given by:

$$\hat{s}^2 = \frac{\hat{\sigma}^2}{n} \qquad (11.13)$$

$\hat{s}$ is known as the standard error of the mean (s.e.o.m.) and is a measure of the uncertainty in q_m.

Given a knowledge of $\hat{s}$ it is possible to calculate values $q_m + \delta_2$ such that the true value of q_m lies between them with a chosen probability P. $q_m + \delta_1$ and $q_m - \delta_2$ are known as the confidence limits and P as the confidence level. For a Normal distribution $\delta_1 = \delta_2 = \delta$ and we have:

$$\delta = t \times \hat{s} \qquad (11.14)$$

The value t (called Student's t) is available in tables, which tabulate t for various values of P and the degrees of freedom, n-1, used in the calculation of s.

Thus for intrinsic errors, the uncertainty will be specified as $\hat{s}$ or as $\pm\delta$ at some specified confidence level.

In the case of influence errors it is usual to specify for an instrument influence error coefficients $(\partial q_i/\partial q_{dn})$ or more usually $(1/q_i).(\partial q_i/\partial q_{dn})$. The actual influence errors, however, depend upon the extent of deviation of the influence quantities from reference conditions. If ε_{dn} is the error due to a deviation of influence quantity q_{dk} by Δq_{dn} from reference conditions, then as shown previously:

$$\varepsilon_{dk} = \frac{\partial q_i}{\partial q_{dn}} \Delta q_{dn} \qquad (11.15)$$

If the actual value of Δq_{dn} is unknown, we may know the p.d.f. of Δq_{dn} : $p_{q_{dn}}(\Delta q_{dn})$. Then the p.d.f. of the influence error due to q_{dk} is:

$$p_{\varepsilon_{dn}}(\varepsilon_{dn}) = \frac{p_{q_{dn}}(\Delta q_{dn})}{\partial q_i / \partial q_d} \qquad (11.16)$$

The mean of $p_{\varepsilon d}(\varepsilon_d)$ is then the best estimate of the influence error ε_d and the variance (or standard deviation) of $p_{\varepsilon d}(\varepsilon_d)$ can then be used as measures of the uncertainty due to influence effects.

Systematic uncertainties are best specified in terms of limits $q_i + \delta_1$ and $q_i - \delta_2$ between which for a given value of q_i the true value of q_m lies. The limits of systematic uncertainty must be estimated by a mathematical analysis of errors.

In accurate measurement it is recommended to specify separately random and systematic uncertainties as described above. In many industrial measurements systematic uncertainties are much larger than random ones and the latter may be neglected.

11.11 ACCURACY, REPEATABILITY, REPRODUCIBILITY

The accuracy of a measuring instrument is the ability of the instrument to give indications equivalent to the true value of the quantity measured.

The quantitative characterisation of accuracy should be given in terms of random and systematic uncertainty. Since the uncertainties will be different at different values in the range of the instrument, the uncertainty should be specified as a function of the quantity measured. In practice it is common to specify the maximum uncertainty in the range, as a fraction of the upper limit of the range. The actual accuracy of the instrument in use depends also on the magnitude of the influence error effects.

In many cases what is of interest is not the accuracy, but the repeatability of an instrument, that is the ability of the instrument to give identical indications or responses, for repeated applications of the same value of the measured quantity under stated conditions of use. It is basically freedom from random intrinsic error. It is expressed quantitatively by a measure of dispersion of the p.d.f. $p(q_i/q_m)$, or generally $\hat{\sigma}$. As with accuracy it should be stated as a function of the measured value.

The accuracy of a measurement is the degree of correspondence of the results of a measurement to the true value of the measured quantity that is freedom from error, (see 2). Accuracy should be stated in terms of the systematic and unsystematic uncertainties.

The repeatability of a measurement is a quantitative expression of the closeness of agreement between successive measurements of the same value of the same quantity carried out by the same method, by the same observer, with the same measuring instruments, at the same location and at appropriately short intervals of time.

The reproducibility of a measurement is an expression of the closeness of the agreement between the results of measurements of the same value of the same quantity after correction for certain errors, where the individual measurements are made under different conditions, say, by different methods, with different measuring instruments and the like.

11.12 INDIRECT MEASUREMENTS

Indirect measurements are measurements in which the value of a quantity is calculated from measurements, made by direct methods, or other quantities linked to the quantity to be measured by a known relationship.

If q_m denotes an indirectly measured quantity, determined from direct measurements of quantities q_1, q_2, ... by the use of the relation:

$$q_m = f(q_1, q_2, \ldots) \quad (11.17)$$

Then the estimate of q_m is taken as:

$$q_m = f(\hat{q}_1, \hat{q}_2, \ldots) \quad (11.18)$$

where $\hat{q}_1$, $\hat{q}_2$, ... are the estimates of the mean of q_1, q_2 ...

Then if for random uncertainties the estimated variances of q_1, q_2, ... are $\hat{\sigma}^2(q_1)$, $\hat{\sigma}^2(q_2)$, ... we have:

$$\hat{\sigma}^2(q_m) = \left(\frac{\partial q_m}{\partial q_1}\right)^2 \hat{\sigma}^2(q_1) + \left(\frac{\partial q_m}{\partial q_2}\right)^2 \hat{\sigma}^2(q_2) + \ldots \quad (11.19)$$

The variance of the mean is given by:

$$\hat{s}^2(\bar{q}_m) = \left(\frac{\partial q_m}{\partial q_1}\right)^2 \hat{s}^2(\bar{q}_i) + \left(\frac{\partial q_m}{\partial q_2}\right)^2 \hat{s}^2(\bar{q}_2) + \ldots \quad (11.20)$$

The above equations neglect higher order terms and also assume that all components are independent of each other.

In the case of systematic uncertainties, $\Delta(q_m)_1$ is the component of systematic uncertainty of q_m due to the systematic uncertainty Δq_1, in q_1, etc. then:

$$\Delta(q_m)_1 = \left|\frac{\partial q_m}{\partial q_1}\right| \Delta q_1 \qquad (11.21)$$

There is no rigorous way of combining the systematic uncertainty components to give the overall systematic uncertainty $\Delta(q_m)$.

Two relations are used in practice. One adds the components:

$$\Delta(q_m) = \left|\frac{\partial q_m}{\partial q_1}\right| \Delta q_1 + \left|\frac{\partial q_m}{\partial q_2}\right| \Delta q_2 + \ldots \quad (11.22)$$

This is likely to be overestimate and represents the estimate of the maximum of uncertainty.

The second method combines the systematic uncertainties in quadrature:

$$\Delta(q_m) = \left\{\left(\frac{\partial q_m}{\partial q_1}\right)^2 (\Delta q_1)^2 + \left(\frac{\partial q_m}{\partial q_2}\right)^2 (\Delta q_2)^2 + \ldots\right\}^{\frac{1}{2}} \quad (11.23)$$

This method tends to underestimate the uncertainty.

There may exist also an uncertainty in the relation f( ). This may be dealt with by introducing them as an uncertain parameter in the above equations.

11.13 ACTIVE MEASUREMENT

In the preceding discussion of errors we considered measurements in which the measured variable acted directly on the measuring instrument. This is the case in the measurement of quantities associated with a power flow as, for example, force, current or temperature.

Quantities which characterise the storage, transformation or transport of energy must be measured by active procedures. In active measurements the magnitude of the measured quantity is determined by interrogating the physical system by a physical input and observing the physical output. The measured quantity is obtained from values of the input and output, using a model of the physical system which relates the measured quantity to the input and output variables.

This is indirect measurement discussed in the preceding section. The errors arise from three basic sources: errors in the determination of the interrogating input quantity, errors in the determination of the observed output quantity and inadequacies of the system model.

BIBLIOGRAPHY

1. Campion, P.J., Burns, J.C., Williams A., 1973, A Code of Practice for the Detailed Statement of Accuracy, London, H.M.S.O.

2. Dietrich C.F., 1973, Uncertainty, Calibration and Probability, London, Hilger.

3. Topping J., 1972, Errors of Observation and their Treatment, London, Chapman and Hall.

4. Hofman D., 1982, Measurement errors, probability and information theory, in P.H.Sydenham, Handbook of Measurement Science, Vol. 1, Theoretical Fundamentals, Chichester, John Wiley & Sons. p.241-275.

This article appears by permission of Pergamon Press Limited, the article being published in their SYSTEMS AND CONTROL ENCYCLOPAEDIA; Thoery, Technology, Applications.

Chapter 12

Metrology

Professor D. J. Whitehouse

12.1 INTRODUCTION

Metrology is the science of measurement and as such is fundamental to science and engineering as well as to everyday life.

Measurement is the description of the properties of objects such as the size or temperature, it is not a description of the object itself. To the student, metrology often appears to be a dull and time consuming subject, in fact it is one of the fastest growing disciplines and is responsible for the dramatic growth in the market penetration in goods of countries which have placed emphasis on quality control.

A fundamental definition (1) is: "Measurement is the process of empirical, objective, assignment of numbers to properties of objects or events in the real world in such a way as to describe them".

Great scientists of the past have commented on its importance. Galileo said "Count what is countable, measure what is measurable and what is not measurable make measurable" Kelvin said "When you can measure what you are speaking about and express it in numbers, you know something about it, but when you cannot measure it, when you cannot express it in numbers, your knowledge is of a meagre and unsatisfactory kind; it may be the beginning of knowledge but you have scarcely in your thoughts advanced to the stage of science".

It is not the intention of this note to go into the formal theory of metrology; this is a subject which is rapidly developing and which is covered elsewhere. This note will concentrate on some of the more general issues. However, the elements of a theory of measurement consist of four basic parts (1).

(1) An objective empirical relational system corresponding to as quality,
(2) a number relational system,
(3) a representation condition,
(4) a uniqueness condition.

The relevance of these four elements will become apparent as the note progresses. Basically these can be brought together by the general statement that a measurement system enables a measure, represented by a number to be obtained of a property which is in itself unique for the object.(1)

One of the prerequisities of a measurement system is that it presupposes that there is some useful property which needs to be measured, so in any situation the first step is to establish the requirement for measurement. For example, one needs to appreciate the importance of hotness before the design of a thermometer is started.

Although the need, subsequent design and application of a measurement system is usually clear and straightforward in the physical sciences such as physics and engineering, it is not quite so simple in sciences such as economics, environmental studies etc., because of their very vagueness and complexity. One such feature of an object which is difficult to quantify is cosmetic appearance for example.

Another important constituent of any general measurement system is its ability to be described by a mathematical model. This step in measurement system theory is becoming more and more important as time goes on as the availability of computing facilities increases and the cost of mechanical development increases.

There are other types of model which can be used, for example the true physical models or the iconic models but these are either too expensive and difficult to model as in the former or less rigorous as in the latter.(1)

A typical block diagram for a measure system is shown in Fig. 12.1 This is not definitive, it just serves to show the type of blocks, inputs and outputs which can be put together to make up a measurement system.

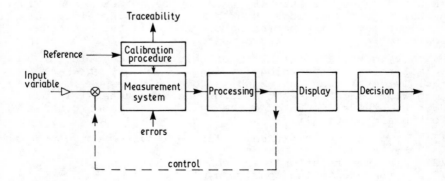

Fig. 12.1

12.2 ELEMENTS OF MEASURING SYSTEM

Fig. 12.1 shows in a simple way some of the important elements in a measuring system. Each of these will be examined briefly in this section, starting basically at the input variable.

12.2.1 The Variable

Measurement information can be conveyed in a number of forms such as energy, mass flow etc., it can be in the form of light or the simple displacement of a mechanical element. For the feature to convey information it has to change either spatially or temporally. Obviously the input variable can be analogue-continuous, discrete continuous or discrete-discontinuous. In all amplitudes can change, frequency, phase or average timing between zero crossings. Sometimes the variable is such that its deterministic characteristics are either not required or impossible to get. In such cases the statistical characteristics of the input would be the relevant parameters. This type of signal is called stochastic. Both deterministic and stochastic signals can be used as test signals to check the overall system.

12.2.2 Reference

The practical embodiment of measurement is the comparison of the feature of an object being measured against a reference or standard. The reference has preferably to have the same physical property as the feature of the object. Also the reference has to be accepted by other investigators if results of measurement are to be communicated. Obviously the reference or standard is man-made conceptually or physically and is meant to be as nearly constant in properties as possible. An obvious example is the unit of length which at present defines the metre in terms of the Krypton 86 radiation line.

Not only does the standard have to be agreed but the procedure for calibrating the measuring system also has to be agreed. As will be seen in the section devoted to errors, these can only be minimised if the measurement procedure images the calibration procedure.

Very often breaking down a measurement requirement which involves the linking of numerous interrelated features just to get one quality is very difficult because of the difficulty of relating the various standards of each feature. In engineering it is usually straightforward but not always.

As an example of how even simple engineering measurements can cause problems, consider the measurement of the out-of-roundness of a manufactured part. This will be used to illustrate a number of points. The measurement feature is not absolute distance but the deviation of the geometry of a part from a true circle. The fundamental or primary standard is still that of length but the working standard is more complicated, it is a reference shape. Although the absolute standard of length as defined above can and will be refined

182 Metrology

with time the basic philosophy remains the same. The options however are larger for the working standard as can be shown. It is possible to derive and use

(1) spatial standards
(2) dynamic temporal standards
(3) mathematical standards.

In the first instance a reference part can be made almost perfectly circular by using the latest manufacturing technology. This can be compared by means of a caliper to the part being measured. Shape differences are measured but not size. This is the direct approach. An indirect measurement is to generate a true dynamic circle in space by means of sophisticated instrumentation, to position this circle near to the part and then to use a caliper to detect differences.

The third way is to realise that the growing importance of mathematical methods in metrology offers considerable advantages over conventional techniques, especially in multi-disciplinary applications such as this.

Nevertheless, it does not involve the blind application of mathematical technique; a degree of knowledge of the nature of the feature to be measured is essential. In this case the requirement is for a knowledge of the spatial bandwidth of the signal.

Consider Fig. 12.2

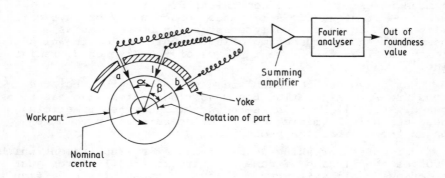

This shows an arrangement in which a part which is truly circular would register a value of out-of-roundness of zero. It comprises of three probes rather than the conventional caliper. Each probe has a different sensitivity 1,a,b - which definitely sounds odd. The probes are at unequal angles α and β and mounted on a yoke. The angles and sensitivities ensure that

$$\left. \begin{array}{l} a \cos \alpha + b \cos \beta - 1 = 0 \\ b \sin \beta - a \sin \alpha = 0 \end{array} \right\}$$

The roundness of any part which is truly round would be

by this arrangement. However, deviations from roundness are seen but in a scrambled form. The out-of-roundness can be extracted, however, by using Fourier Analysis. It can be shown (2) that if a harmonic analysis is made of the combined signal $s(\theta)$ emerging from the summing amplifier such that

$$s(\theta) = \sum c'_n e^{j(n\theta + \psi'_n)}$$

and this is compared with a similar harmonic analysis of the true out of roundness signal $f(\theta)$
where
$$f(\theta) = \sum c_n e^{j(n\theta + \psi_n)}$$

then in order to convert C_n' to C_n and ψ_n' to ψ_n the C_n' amplitudes have to be modified by a factor k.

Where $k^2 = [(a\cos\alpha + b\cos n\beta - 1)^2 + (b\sin n\beta - a\sin n\alpha)^2]^{\frac{1}{2}}$

and the phases ψ_n' modified by a factor γ

where $\gamma = \left[\dfrac{b\sin n\beta - a\sin n\alpha}{a\cos n\alpha + b\cos n\beta - 1} \right]$

Using this method the out-of-roundness can be evaluated despite the fact that the measuring device has an axis of imperfect rotation.

In effect a mathematical reference has been derived. This is just one example of the increasing use of mathematics in metrology.

12.2.3 Calibration

In order that devices and systems can be interchangeable all measuring systems have to be calibrated and the physical attribute of which the calibration is made such as length, mass temperature or whatever has to be carried out using units which can be directly traced to the International S.I. standards of which there are seven. Mass (kg), temperature (ok), current (A), luminous intensity (cd), amount of substance (mol), length (m) and time (s). From these all other units such as the Newton of force can be derived. The seven basic standards are called base units and they are considered to be independent of each other.

12.2.4 Measurement procedure (3)

There are a number of ways of making a measurement. The "direct" method is one in which the value of a quantity to be measured is obtained directly. In a direct measurement the unit of measurement has to be the same as the scale of reference used.

Indirect measurement is that in which the parameter sought is obtained by use of intermediate stages of different units which are correlated to each other.

The comparison method is based on the comparison

of the quantity to be measured with a known value of the same quantity or with a known value of another quantity which is a function of the quantity to be measured.

The differential method comparison gives a slight difference between the standard and the measurand that is used to apportion a small additive value to the standard value. The null method is similar but not identical because in the latter the difference between the standard and the measurand is adjusted to be zero - the indicator need not be calibrated!

There are other variants on these methods but the ones mentioned are the most often used.

12.3 MEASUREMENT TERMS AND PERFORMANCE OF INSTRUMENT SYSTEM

There are three or four concepts which are used when conducting a measurement or describing the performance of an instrument.

Resolution is the term which relates to the ability of an instrument to respond to a small change in the measured quantity. It is sometimes called discrimination. It does not relate to accuracy.

One big fault often met with in scientific reports is the confusion between resolution, repeatability, reproducibility and accuracy. Sometimes the resolution is written in such a way that it could be misconstrued as the accuracy. The accuracy formally is that quality which characterises the ability of a measuring instrument to give indications approximating to the true value of the measured quantity. The repeatibility or sometimes the precision is that quality which characterizes the ability of a measuring instrument to give the same value as the quantity measured. Precision refers to the same piece of apparatus. If the same parameter is measured over a long term using different people with different apparatus the quality is called the reproducibility of the parameter measurement.

Another term used in instrument performance is linearity. This again is not accuracy. Linearity only expresses how values lie on a proportional scale. The extent of linearity is sometimes called the range. Very often the quality of design of instrument is linked to the ratio of the range divided by the resolution. Fig. 12.3 shows a composite scale of typical transducer systems indicating how with a time factor incorporated into the merit order a good idea of relative cost of sensing can be obtained. Obviously instead of frequency response, response time could have been used. Other criteria should be included in a comprehensive system of classifying instrument systems. Most often these are concerned with the non-linearities in the system of which there are more than one form. For example, explicit non-linearities imply that the output is determined explicitly from the input value - although not in a linear way. Implicit non-linearity is not so clear. The output may be dependent

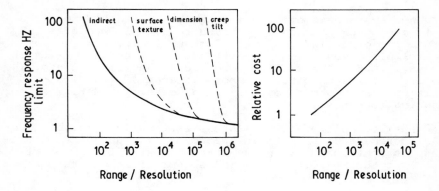

Fig. 12.3   Performance of Measurement Systems

also on the derivatives of the input. In the explicit form
single valued non-linearities occur as in saturation and
dead band multivalued non-linearities also occur as in
hysteresis. For a complete characterization of the system
performance it would normally be in order to specify the
transfer of the measuring system covering the operational
frequency range and encompassing the linear amplitude. There
are of course, numerous ways of dealing with the non-linear
aspects of the characteristics such as the describing function
method, phase plane analysis and the relatively recent state
variable approach. It is necessary to mention that the
complete characterization includes mechanical elements,
interfaces, amplifiers filters and recorder responses.

12.4   ERRORS

The theory of measurement is not just a discussion of
procedures and instruments it is equally concerned with the
identification, evaluation, and understanding of errors.
It is the error in measurement which determines the uncertainty of the result and hence the ultimate accuracy and
repeatibility.

12.4.1   Types of error

Systematic error can be caused by errors in measurement
procedure, errors in the standard used for calibration, errors
in the device used for measurement, errors even in the way the
operator performs and bias errors in the environment. By
definition systematic error can be determined and eliminated.
However, the real problem with systematic error is realising
that it exists. Its presence cannot be revealed by mere
repetition as can random error which makes it very difficult

to deal with unless the same object attribute is measured by an independent method.

Once systematic error has been identified and measured (2) it can often be reduced or even removed in its effect by storing the error in a computer look-up table. By the same token simple cases of nonlinearity can also be dealt with in this way. Often under these conditions the limiting usefulness of an instrument becomes not its accuracy but its precision. Such a case exists in a metrology instrument for measuring roundness.

As an example of how to assess systematic error consider the measurement of an unknown specimen $f(\theta)$ using an out-of-roundness instrument having an unknown systematic error of $s(\theta)$. The output $g(\theta)$ from the sensor after one revolution is $g_1(\theta) = f(\theta) + s(\theta)$.

Then the component is shifted by π and a second result obtained

$$g_2(\theta) = f(\theta + \pi) + s(\theta) = -f(\theta) + s(\theta)$$

from which $s(\theta) = (g_1(\theta) + g_2(\theta))/2$ by the same token, at the same time $f(\theta)$ can be evaluated as $(g_1(\theta) - g_2(\theta))/2$.

12.4.2 Random errors

These are the value of the uncertainty of repetitive measurements taken as an instrument. These errors can be caused by the environment, vibration instrument noise and many other sources. Most often they are mechanical or electrical in nature. Depending on the bandwidth of the signal to be measured the random noise can often be integrated out of the result by a low pass filter or integrator. Simple tricks can be used to remove or at least reduce mechanical random error by means of using differential gauging as shown in Fig. 12.4. At one time such a technique would have been difficult to apply because the signal to be measured, in this case the step, becomes scrambled. However, with the advent of cheap computing the true signal can be easily extracted.

12.4.3 Error combinations

Errors can manifest themselves in a number of ways. One example corresponding to an additive error is a zero position shift. Another corresponding to a sensitivity change is multiplicative.

12.4.4 Error models

Systematic errors are often specified in terms of the transfer function of the measurement system. For example, if the true transfer function is $F(s) = N(s)/M(s)$ and the measured transfer function is $F_1(s) = N_1(s)/M_1(s)$ the measurement error transfer function is described as

Metrology 187

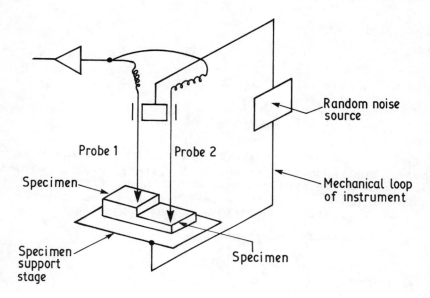

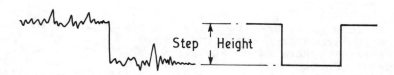

SINGLE PROBE TRACE   DIFFERENTIAL PROBE TRACE

Fig. 12.4   Reducing Effect of Random Noise

$$F_E(p) = \frac{F_1(s)}{F(s)} - 1$$

Random errors have to be described probabilistically in terms of the amplitude probability density function of the random signal and the parameters used to describe it such as the mean, standard deviation and in some cases the characteristic function of the distribution.

12.4.5   Error propogation

Errors are usually assumed to be small relative to the quantity being measured and as such their deviation from the true value can be expanded in a Taylor's expansion.

If the desired quantity y is a function of a number of different measured values x_1 x_2 x_3 x_4 ...

such that $y = f(x_1 x_2 x_3 \ldots x_4)$

then the systematic indirect measurement error of y which

is δy is given by

$$\delta y = \sum_{i=1}^{n} \frac{\partial y}{\partial x_i} \delta x_i$$

Using logarithmic differentiation

$$\frac{\delta y}{y} = \sum_{i=1}^{n} \frac{\partial \ln(y)}{\partial x_i} \delta x_i$$

For values x_1 etc. which are correlated this formula does not hold. For random errors specific systematic error values like δx_i also are meaningless. Under these circumstances the standard deviation of the errors have to be used. Even this is not complete if the errors are non-Gaussian.

The corresponding equation

$$S_e = \pm \left[\sum_{i=1}^{n} \left(\frac{\partial y}{\partial x_i} \right) S_i^2 \right]^{\frac{1}{2}}$$

12.5 PROCESSING

In some cases despite every precaution it is impossible to remove errors from the signal measured. The errors could be referred to here as noise. An example might well be in electronic Shott noise or Johnson noise. Under these circumstances it is important that there are techniques available which can extract the maximum amount of useful signal from what has been measured. This problem basically separates itself into two categories; that of identifying the presence or absence of detail and that of actually quantifying the feature or parameter. These two branches of processing are called loosely pattern recognition and parameter estimation. These subjects will be dealt with in some detail elsewhere and will therefore be omitted in this paper.

The basic importance of this part of the metrology system is that any control of the overall process will be initiated by this output whether it is direct by means of closed loop control as shown in Fig 12.1 or indirectly by means of operator intervention based upon decisions taken.

12.6 RECENT TRENDS IN METROLOGY

There are three general trends developing in modern metrology; they are:

12.6.1 the increasing use of computers in data storage, data processing, data transmission and display

This has resulted in

12.6.1.1 an increase in effective accuracy by storing systematic errors

12.6.1.2 the measurement of more derived parameters especially in engineering metrology and in particular surface and physical metrology such as best fit and minimum zone cylinders

12.6.1.3 less ambiguity in reading and interpreting displays with the consequent decrease in measurement uncertainty.

12.6.1.4 More straightforward automatic control of measuring instruments.

12.6.1.5 The use of CAD is now enabling better design of equipment. The resultant computer models of transducers, sensors and electrical systems has meant that devices are becoming cheaper and more efficient.

12.6.1.6 Semiconductor material technology is being incorporated into sensor design with the result that measuring systems will become smaller and therefore able to be located where the measurement of critical parameters is needed.

12.6.2 The increasing use of statistical techniques, especially random process analysis in processing the data.

This has resulted in

12.6.2.1 a greater understanding of which parameters need to be measured in order to control a functional situation.

12.6.2.2 An extension of the useful range of existing sensors and transducers by extracting the last fraction of information from the available data.

12.6.3 An increasing use of coherent optics in measurement, this has resulted in:

12.6.3.1 a growing interest in non-contacting mechanical measuring systems

12.6.3.2 much faster data processing capability thereby opening up real time control of systems in industry.

12.7 CONCLUSIONS

Metrology and through it quality control is showing its teeth after an inauspicious start. The signs are that it will grow in importance.

12.8 REFERENCES

1. Finkelstein L., 1982, 'Theory and Philosophy of Measurement'. Handbook of Measurement Science, 1, p.1, Wiley.

2. Whitehouse D.J., 1976, 'Some theoretical aspects of error separation techniques in surface metrology.' J.Phys.E.Scientific Instruments, 9.

3. Sydenham P.H., 1982, 'Measurements Models and Systems.' Handbook of Measurement Science,' 1, Wiley.

Chapter 13

Automatic inspection

Dr. W. J. Hill

13.1 INTRODUCTION

'Inspection' is used in this paper to mean the examination of a product for departure from acceptable functional properties. This is simply the determination, by whatever means, of whether a material or part will satisfactorily perform its intended function. This includes, for example, the inspection of structural components, such as beams and ropes, for faults and weaknesses, the inspection of glass bottles for distortion of shape, the examination of sheet metal for surface defects such as stains, scratches, rust spots etc. Such inspection, in manufacture or in service, is an important, common and general requirement. Defects may arise at virtually any stage of a production process, and subsequently in service. It is rarely, if ever, possible to guarantee consistently satisfactory functional properties by controlling the manufacturing process. Inspection therefore serves a sorting function, allowing safety and other standards to be maintained without excessive demands on production and associated economic costs.

Our attention will be restricted to those methods of inspection which do not impair the future usefulness of the product. This is traditionally the area of Non-destructive Testing (NDT). We might define <u>Automatic</u> Inspection as NDT without human intervention. This leads to our focus of interest being somewhat different from traditional texts on NDT (McGonnagle (1), Egerton (2), for example). These tend to concentrate on the wide variety of methods available for interrogating the inspected item, so as to elucidate the physical and chemical principles on which these methods are based, and to provide guidance for their application and interpretation of the results. This interpretation and the consequent decision to accept or reject (or repair) the product is usually left to trained and experienced human operators. The most difficult task in automating the inspection process lies in this stage of interpretation and decision making. In general this presents major difficulties of both principle and practice, and demands the use of the most sophisticated and powerful methods currently available for the processing of information. Even then, in many cases, significant extensions are required of these methods, and the most difficult tasks demand major conceptual advances in the field.

13.2 GENERAL PRINCIPLES

Most methods of NDT rely on indirect measurement. They do not measure directly the functional properties of interest, but instead measure physical or chemical properties which may be related (more or less closely) to those functional properties. In radiography, for example, the transmission of X-ray or gamma radiation through a test specimen is recorded. This transmission is directly affected by local thickness and density variations in the specimen. The recorded pattern is therefore a direct measure of such variations. For the detection and characterisation of defects in welded joints, pipes, castings etc. it is necessary to relate such variations to the defects themselves. This is an a-priori requirement essential to the correct interpretation of the test results. It is further necessary to relate the defects to the functional properties of interest, such as the mechanical properties of the specimen. As another example, eddy current techniques can be used to inspect electrically conducting specimens for defects, irregularities in structure, and variations in composition. The quantity directly measured is the electrical impedance of the exciting coil held in close proximity to the specimen (or else the induced emf in a second pick-up coil). Variations of impedance can be related to structural characteristics of the specimen, and thus to functional properties such as mechanical strength. This relationship however may be exceedingly complex, rendering the task of interpretation difficult and problematic - particularly if the test is to be fully automated.

In general, the methods of NDT can be characterised by five distinct stages:

(i) INTERROGATION - the test specimen is interrogated by an energy input.
(ii) MODULATION - the specimen interacts with the interrogating energy, varying its characteristics, and thus modulating that energy with useful information.
(iii) SENSING - the modulation is sensed and recorded in a form suitable for subsequent analysis.
(iv) PERCEPTION - the modulation record is analysed so as to deduce the relevant characteristics of the specimen.
(v) DECISION-MAKING - the perceived characteristics are related to the relevant functional properties of the specimen, and a decision is made to accept, reject, or repair the specimen.

These five stages are clearly evident in the two examples of radiography and eddy current testing already discussed. Table 1 identifies this structure for some other common methods of NDT. Note the reliance upon the skills and capabilities of human beings in stages (iv) and (v).

13.3 THE RELEVANCE OF MODELS

It is clear that the most difficult and critical problems arise in the areas of perception and decision-making. Perception, the process of relating sensed signals to the relevant characteristics of the specimen, is difficult when a suitable model of the modulation process is not available. This may be because the process is not sufficiently well understood, or because the process is so complicated that a suitable model cannot be devised. If a mathematical model is available, expressing the relation between the modulation signal or pattern and the relevant geometrical and structural properties of the specimen, then it should be possible (in principle, at least) to deduce these properties by an on-line system identification or parameter estimation procedure. Investigations have shown, however, that in general given a model non-linear in the parameters, and a response signal contaminated by noise, identification and parameter estimation is difficult and prone to error.

In the area of decision-making, the problem is to relate perceived properties of the specimen to functional properties of interest. Such relationships can be exceedingly complex, with many inter-dependent factors, and suitable models are generally not available. Decisions therefore must often rely on human judgement. Ideally, all defects should be describable by a small number of independent and readily perceivable parameters, and the design staff should specify acceptable and unacceptable combinations of these parameters. This is rarely the case, and the resulting 'grey areas' must be dealt with by the judgement and skill of trained operators, often in collaboration with the designer.

13.4. OPTICAL SURFACE INSPECTION

As a concrete example of a problem in automatic inspection that has received considerable research attention in recent years, we may consider the automatic optical surface inspection of cold-rolled steel strip (Hill et al (3)). The problem is to fully inspect the steel surface for defects, as the strip emerges from the temper mill, with strip speeds of up to 25 m s^{-1}. Defects on the steel surface, such as surface lamination, scale, pits, rust, stains, etc., must be detected and identified with a high degree of confidence. This will allow the steel to be sorted for different customers, and also permit adjustments to be made to the production process so as to minimise the incidence of defects. Some 50 different classes of defect must ultimately be distinguished, and for the smallest the inspection system must achieve a resolution of about 1 mm at the inspected surface.

Many steel producers throughout the world have tackled this problem, each adopting a somewhat different strategy for the research. In a research programme at the Measurement and Instrumentation Centre of The City University, the strategy adopted has been to investigate the problem via computer simulation, but processing 'real' data from defects. In describing this research the five distinct stages of the inspection process will be treated separately, with emphasis on the problems of perception in an automated system.

13.4.1 Interrogation.

The steel surface is interrogated optically with a high-intensity beam of laser light, via a high-speed scanner of the form shown in Figure 13.1 This is a flying-spot system developed by SIRA Ltd., and provides three simultaneous views of the surface. A 5mw Helium-Neon laser provides the interrogating energy, and the beam is reflected from a multi-faceted rotating prism, spinning at 400 revolutions per second. The beam is focussed onto the steel surface so that each facet of the prism as it rotates, scans the beam across the full width of the material (up to 1.5m). With proper choice of scan-rate and spot size, complete coverage of the surface is achieved (in the fashion of a raster scan) as the steel surface passes beneath the scanner. For this research, the beam was focussed to a spot of 1mm diameter at the surface. The laser is used simply as a convenient source of a high intensity collimated beam of light. The coherence of the light is not exploited in this application.

13.4.2 Modulation.

The steel surface interacts with the interrogating energy by deflecting, scattering and absorbing the incident light - in accordance with the specific characteristics of the surface at the point of interrogation. Thus, as the beam scans across the surface, a scatter pattern is generated which changes continously, and carries information at each instant about the surface properties of interest.

13.4.3 Sensing.

Three photomultipliers positioned in the scattered light field provide a partial and selective characterisation of the scatter pattern, and its evolution in time as the interrogating beam scans across the surface. Early experiments showed that three such 'views' of the surface were required to adequately sense the full range of significant surface defects. The photomultipliers are positioned at the specular angle (No. 1), 7 off-specular (No 2), and 30 off-specular (No 3). 75mm photomultipliers are used for fast response, and each of the three signal channels requires a bandwidth approaching 25MHz at the highest strip speeds. For the research program, the signals were digitised and recorded onto magnetic tape for subsequent computer analysis. Figures 13.2, 13.3 and 13.4 show typical sensed data over a 30 cm square sample of steel.

Automatic inspection 195

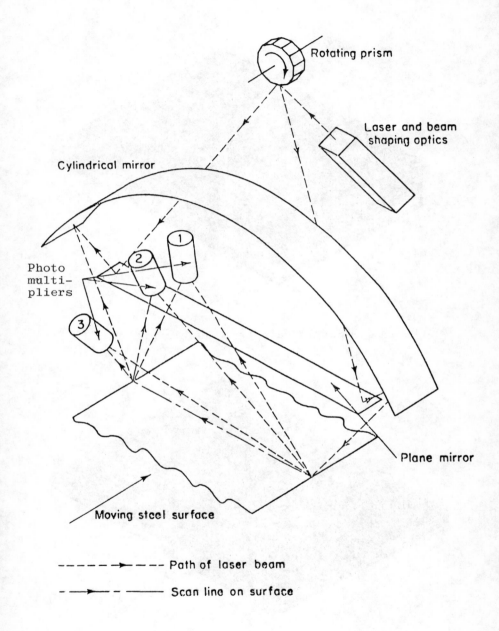

Figure 13.1 - Laser Scanner schematic (SIRA Limited)

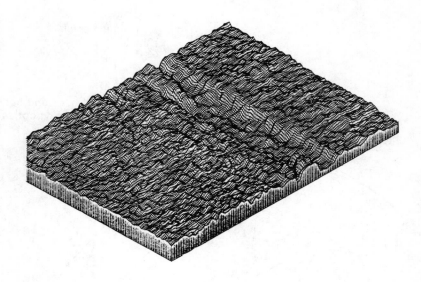

Figure 13.2. - Sensed data from the first photomultiplier

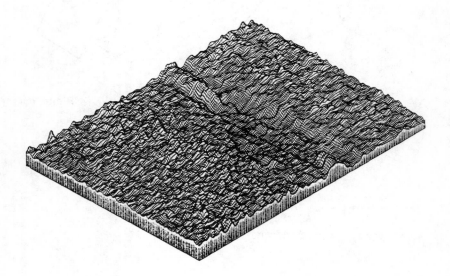

Figure 13.3 Sensed data from the second photomultiplier (7 degrees of-specular).

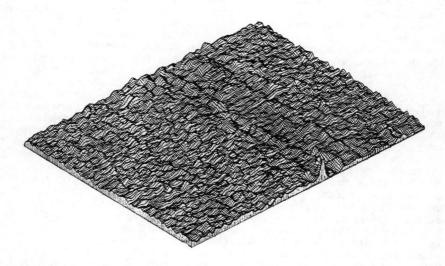

Figure 13.4 Sensed data from the third photomultiplier (30 degrees off-specular)

13.4.4 Perception.

For perception, the sensed data must be analysed so as to locate and identify surface defects. Ideally this analysis would be guided by suitable models relating each defect type to characteristic surface properties, such as profile, reflectivity and colour, and by a model relating these properties to the resulting scatter pattern, and hence to the signal on each photomultiplier channel. It would then be possible, in principle, to solve the 'inverse scattering' problem, so as to relate signal parameters to particular surface properties, and thus to the presence or absence of specific defects. This would provide a complete solution to the problem of perception.

In fact, it is not currently possible even to define the surface properties which characterise each defect class. The classes are <u>defined</u> by their origin. Skin lamination, for example, orignates from a bubble of air trapped within the steel ingot, which migrates to the surface during rolling, and eventually leads to a thin strip of the surface being torn away. The different types of defect can be <u>recognised</u> from an examination of the surface by a trained and experienced human inspector, with a high degree of reliability, but the identification proceedure is not well understood, and extremely difficult to elucidate.

By contrast, modelling the relationship between surface properties and the resulting scatter pattern (and hence to the signal on each photomultiplier channel) is an easier task. Models have been developed by treating the surface as an assembly of small mirrors, positioned to match the surface profile. More sophisticated modelling allows for diffraction and scattering effects via the Kirchoff theory of electromagnetic scattering (Obray(4)). Agreement between the predictions of these models and observed scatter patterns from different surfaces has been demonstrated, but primarily in terms of the statistical properties of both. This holds promise for the remote characterisation of surface texture, but is not necessarily useful for the detection and identification of surface defects. Work is continuing in this area.

13.4.4.1 Automatic Pattern Recognition. In the absence of suitable models relating defect type to surface properties, and surface properties to the signals on each channel, the methods of automatic pattern recognition have been used to establish 'learned' correspondences between channel signals and defect type. This is equivalent to a 'black-box' model which accepts as inputs a set of parameters derived from the channel signals, and responds with the name of the defect type most likely to have generated such signals. The essential characteristic of this approach is the ability to 'learn by observation' the required correspondences, without the need for a detailed model of the signal generating process. For example, if defect type A tends always to produce signals of high amplitude, and defect type B signals of low amplitude, this fact can be discovered simply by examining a large number of signals from each of the two types of defect, and subsequently used to separate them. It is not necessary to know why these particular defect types produce signals with those characteristics.

In practice, of course, the correspondence between signal parameters and defect types is far more complex than this. For steel surface defects, it turns out that around 10 different parameters must be processed simultaneously to discover and exploit the interrelated variations and dependencies between them, which are characteristic of the different defect types. Furthermore, correnposdences are likely to be statistical rather than deterministic, with a particular defect type tending to influence the signals in a particular way.

The methods of automatic pattern recognition embrace such problems. The signal parameters are known as 'features' of the defect, and the set of parameters taken together define the co-ordinates of a multi-dimensional 'feature space'. The system is presented with examples of each defect type, and by noting the location in the feature space of each example, the necessary correspondences are 'learned'. This amounts to establishing a partition of the space, with each region assigned to a particular defect type. Figure 13.5 illustrates this in two dimensions (see Duda and Hart (5) for more information on feature space techniques).

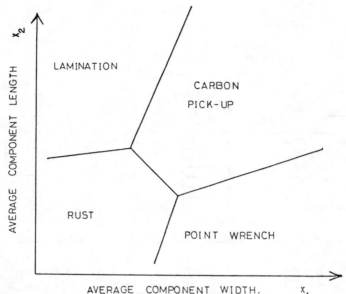

Figure 13.5 A two-dimensional feature space

13.4.4.2 <u>Signal detection, delineation, and parameterisation</u>. Feature space techniques assume prior parameterisation of the signals. This in turn implies that each defect has been detected and delineated (i.e. the defect boundary located) in amongst the general background noise arising from innocuous surface structure, electronic noise etc. In order to adequately capture the structure of each defect, detection and delineation must be organised in a three level hierachy. The first level has the task of processing each line scan, on all three signal channels, so as to identify the location and extent of all defects crossed by that line scan. The second level must associate these signals between scans, so as to locate compact clusters arising from the same defect component. The third and final level must associate defect components belonging to the same defect. A defect component is typically, a single rust spot, whereas the complete defect consists of many such spots. An adequate parameterisation of such defects requires that the separate identity of each component be preserved within the delineation process. Figure 13.6 illustrates this three level hierachy.

In this research, signal detection and the first level of delination is achieved with a bank of filters, each matched to an archetypal defect signal, so that their combination spans the range of signals of interest. The second level of delineation is based upon the proximity and similarity of the signals detected at the first level. Signals are required to be typically less than 1.5 mm apart along each scan, and less than 3 mm apart across scans, as well as being on the same signal channel and similar in waveform, in order for them to be associated with the same defect component. The third level of delineation is organised similarly to the second, but with proximity thresholds ten times larger (i.e. 15 mm along scans, 30 mm across scans). Because of the very high

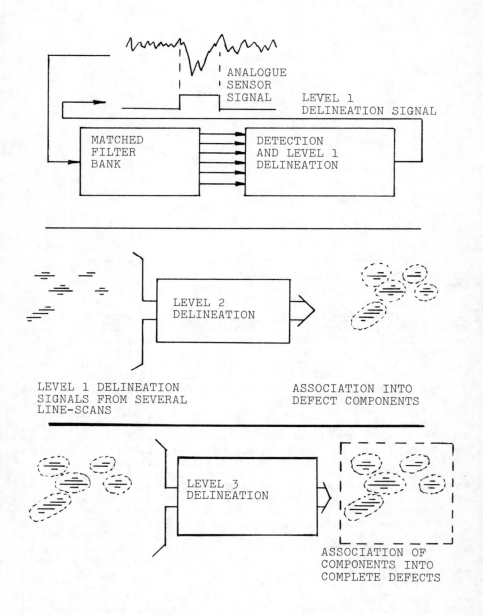

Figure 13.6 - The processing hierarchy for defect delineation

data rates on each signal channel all of this early processing operates on a 'scan-by-scan' basis, with minimal storage of scan data.

Parameterisation proceeds hand-in-hand with the three levels of delineation. At the first level, parameters are extracted from each line scan, as illustrated in Figure 13.7 As these signals are associated at the second level of delination, the first level parameters are combined and extended to form a parameterisation of each defect component. Thus, for example, 'signal amplitude' parameters extracted at the first level are combined to give the second level parameter 'mean signal amplitude' for each defect component. Other second level parameters will characterise the size and shape of each component (width, length, area, etc.).

Similarly, as defect components are associated at the third level of processing, to form complete defects, so the component parameters are combined and extended to form a complete parameterisation of the whole defect. This includes such parameters as the number of components, average component area, variance of component area, overall width and length of the defect, etc. This final parameterisation is rich in detail reflecting both the fine optical structure of each defect, as well as more global characteristics such as overall width and length.

With this parameter set, and the pattern recognition techniques already described, it has proved possible to correctly identify 85% of the defects in a set of test data with the following composition:

DEFECT TYPE	NUMBER OF SAMPLES
Skin Lamination	20
Slivers	17
Fleck Scale	26
Scrap Marks	10
Pick Up	15
Carbon Pick Up	7
Coil Digs	21
Rust Spots	49
TOTAL	165

Although an on-line instrument would need to deal with a wider range of defect types, this level of performance constitutes a satisfactory solution to the perception problem in a fully automated system.

13.4.5 Decision Making.

For steel inspection, the ultimate aim is to reach a set of decisions on the functional properties of the steel surface. This requires that defect type, severity and spatial distribution be related to the needs of different parameters of the production process. This is a suitable area for the newly emerging techniques of 'Expert Systems', whereby the accumulated knowledge and experience of human inspectors could be embodied into a rule-based system capable of reaching such decisions. Little effort has so far been devoted to this task, because it is not yet necessary to automate it. An on-line inspection system could produce a record of defect type, location and extent, for subsequent analysis by the human operator. Thus, the automatic system could automate the process of interrogation, sensing and perception, and leave the decision-making process to the human being. This would

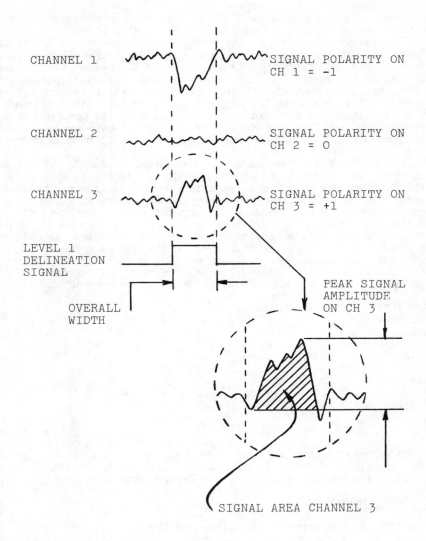

Figure 13.7 Level 1 defect parameters

consitiute a major advance on current practice, with very considerable economic benefits.

REFERENCES
1. McGonnagle, W.J., 1971, 'Nondestructive Testing', Gordon and Breach, New York, U.S.A.

2. Egerton, H.B., 1969, 'Nondestructive Testing', Oxford University Press, London, England.

3. Hill, W.J., 1983, 'Signal Processing for Automatic Optical Surface Inspection of Steel Strip', Transactions of the Institute of Measurement and Control (to appear in 1983).

4. Obray, C.D., 1979, 'Distributions of the Intensity of the Scattered Field from a Rough Surface', Research Memorandum No DSS/CDO/180, Department of Systems Science, The City University, London, England.

5. Duda, R.O., Hart, P.E., 1973, 'Pattern Classification and Scene Analysis', Wiley, New York, U.S.A.

204 Automatic inspection

TABLE 1A Methods of Non-destructive Testing

METHOD	INTERROGATING ENERGY	MODULATION PROCESS	SENSING	PERCEPTION	DECISION MAKING
1. Optical (visual)	Visible light	Reflection, scattering, absorption, shadowing	Eye, film, VTR etc.	Trained operator	Trained operator
2. Radiography	X-rays, Gamma rays, Neutron beam	Attenuation, scattering	Film, fluorescent screen, VTR etc.	Trained operator	Trained operator
3. Ultrasonics	Ultrasound in the form of stress waves	Reflection, scattering at discontinuities	Piezo-electric crystal, record of electrical signal	Trained operator or electronic circuitry (thickness measurement)	Trained operator or electronic circuitry
4. Magnetic particle flaw detection	Magnetic field	Field shaping by structure and material properties o specimen	Film of magnetic particles, with distribution recorded on film, VTR etc.	Trained operator	Trained operator
5. Eddy current inspection	Alternating magnetic field	Discontinuities and inhomogeneities affect pattern and strength of induced eddy currents	Impedance variations of exciting coil, or induced EMF in pick-up coil. Recorded graphically, digitally etc.	Trained operator comparison against standards	Trained operator

TABLE 1B Methods of Non-destructive Testing

METHOD	INTERROGATING ENERGY	MODULATION PROCESS	SENSING	PERCEPTION	DECISION MAKING
6. Acoustic emission	Mechanical stress	Defect growth generating stress waves	Piezo-electric transducers. Signal recorded graphically, digitally etc.	Detection and location of defect growth, time history etc.	Trained operator. Computer models etc.
7. Holography	Coherent light	Reflection, scattering at surface of specimen	Photographic plate	Reconstruction and visual examination, mensuration, comparison.	Trained operator

Chapter 14

Introduction to analytical measurement

Professor G. F. Kirkbright

14.1 INTRODUCTION

14.2 Concepts in Measurement of Chemical Parameters

The measurement of chemical quantities may be <u>qualitative</u> or <u>quantitative</u> according to the need to know the nature of the material under examination or the mass and/or concentration of a particular component of a material of interest. Traditionally there are two types of analytical technique:

(a) <u>Absolute Methods</u> In which quantitative analytical data may be obtained (usually by classical means) concerning the mass or concentration of a particular component in a sample. This is usually established directly employing the stoichiometry of the reactions involved and knowledge of the atomic or molecular weight and molecular formula of the species of concern. The principal classical methods of absolute analysis involve the determination of weight (gravimetric analysis) or volume (volumetric analysis).

(b) <u>Empirical Methods</u> Most analytical measurements, particularly those which employ instrumentation, are empirical in nature and employ arbitrary "signals" or non-stoichiometric reactions recorded instrumentally and it is necessary under these circumstances to relate the observed signal or extent of reaction for the material studied to that produced by a known amount of a substance (standard). It is possible to illustrate some usable but non-stoichiometric solution reactions (the formation of weak complexes used in titrations) and to illustrate the same effect with all instrumental methods (for example, polarography, flame photometry, X-ray fluorescence, nuclear magnetic resonance and mass spectrometry).

14.3 Sensitivity

The definition of sensitivity for analytical reactions and the response of analytical instrumentation is defined for the particular analytical method concerned as the differential quotient dS/dC of the "signal" or observed variable with respect to concentration or mass of material to be examined (or determined), i.e. the slope of the growth curve or "calibration graph". Essentially the differential quotient defined in this way frequently governs the attainable precision within the range of mass/concentration examinable.

14.4 Accuracy

The accuracy of a measurement made with an instrumentation system, whether this be in the laboratory or on-line, is defined as the nearness of the indicated calibrated value to the "true" value. Thus it is necessary to consider in all instrumentation measurements applied for any parameter (whether it be chemical or physical parameter which is transduced) the systematic errors which may occur in the measurement. The source of these systematic errors may originate from the existence of "blanks" due to spurious contamination of the sample examined, instrumental offsets or poor calibration, or variation in the local environmental conditions (for example, temperature or pressure) which affect the nature of the sample, the transducer or the complete instrumentation system. The definition of the "accuracy" of an instrumentation system used for any measurement is therefore a complex problem. Frequent calibration of the apparatus employed and a precise definition of the nature of the sample (using a standard flow stream or material to be examined) is thus required. Even for standard reference materials there is frequently no knowledge of the absolute value of the concentration or mass of the component in that material; values previously obtained to establish these "standards" represent frequently only the mean of many values obtained by different laboratories by a range of different measurement techniques. Most instrumental methods of analysis, particularly for the characterisation of chemical parameters, require the availability of these standard reference materials. This has led to considerable interest in standardisation programmes via national bodies world-wide (for example, National Bureau of Standards, International Standards Organisation, BCR (Europe) etc.).

14.4 Precision

It is important not to confuse the terms "precision" and "accuracy" in consideration of any measurement. As defined above accuracy must be used in relation to the definition of possible systematic errors whereas "precision" is to be applied in relation to the random errors in any measurement. Under truly random error conditions the normal law of error can be applied which predicts a Gaussian distribution of the frequency of occurrence of observations. The relative standard deviation in a series of measurements is used to evaluate the magnitude of the random errors observed. As will be apparent from Figure 1, it is possible to have measurements at high precision which are inaccurate; when precision is low, however, accurate measurements are difficult to obtain.

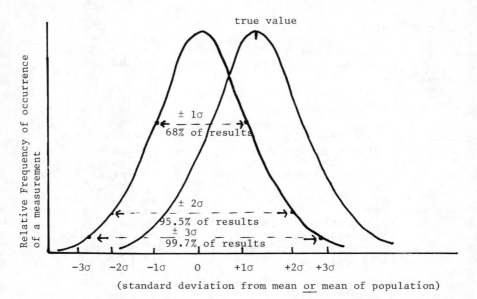

Figure 1. Relative frequency of occurrence of a measurement vs. standard deviation from the mean (or mean of population).

From a consideration of the theory of measurements which have only random error (no systematic errors) it can be shown that most observations will fall within ± 3 standard deviation units of the mean. The vertical axis in Figure 1 represents the relative frequency of the currents over measurement (or its error). The horizontal axis is divided into units of standard deviation, σ, of the population. The curve displays three distinct properties. (a) Because the curve is normal or symmetrical there tends to be a negative error for every positive error of the same absolute value. (b) The relative frequency of measurements having a small area is very great. Over 68% of the measurements fall within ± 1σ unit of the mean. (c) The relative frequency of measurements having a large area is very small. Because 99.7% of the measurements fall within ± 3σ units of the mean, only 0.26% of the measurements fall outside these limits. These latter measurements may then be said to have a large error and there are obviously only a few of them.

A knowledge of the normal statistical distribution of measurements and their errors is useful in evaluating large numbers of measurements. For example, each week in an industrial process control laboratory many hundreds of samples of single cell protein product are analysed for water and fat content using an automated analytical instrument. To be sure that the instrument functions correctly every tenth sample is a standard sample of known water and fat content. When aliquots of the same standard sample have been analysed many times the standard deviation for observations of water and fat content in this standard sample is calculated.If it is assumed that the standard deviation of the unknowns is the same as for the standard samples, the laboratory will know that the analysis of approximately two thirds of the samples can be reproduced within ± 1 standard deviation unit and 95% can be reproduced within ± 2 standard deviation units, etc. Thus the laboratory has a good indication of the degree of confidence that can be placed in their fat and water analyses on the product from the process line. The possibility of systematic error in the actual content of water and fat in the standard sample cannot still, however, be precluded.

14.6 Limits of Detection

It is frequently necessary to assess the utility of instrumental analytical techniques for the determination of various components on the basis of the precision with which analytical observations may be made. At low concentrations one criterion which may be employed is that of <u>detection limit</u>. The detection limit would then define the minimum mass or concentration of the component under investigation whose presence could be monitored by the technique involved. Naturally at low concentrations the detectability of a species is directly controlled by the precision attainable in its measurement (random error) via the signal-to-noise ratio obtained with the instrumental arrangement employed. It would be normal to use two or three times the relative standard deviation (random error) in the measurement depending on the confidence limits required. As shown in Figure 2 the detection limit is commonly defined as that concentration or mass of component which produces a signal-to-noise ratio of 2 in the monitoring system (i.e. a coefficient of variation of 50% corresponding to a relative standard deviation of 0.5). Figure 2 shows the manner typically in which precision degrades as the detection limit is approached and the signal-to-noise ratio degrades. In contrast to the "sensitivity" which reflects the slope of the analytical growth curve and reflects the fundamental capabilities of the technique employed to provide analytical information, the detection limit reflects more frequently on the skill with which the system has been assembled and its particular performance characteristics. Thus two instruments of identical construction might give rise to quite different detection limits in operation in the analysis of a particular component in a flow stream or in a laboratory environment.

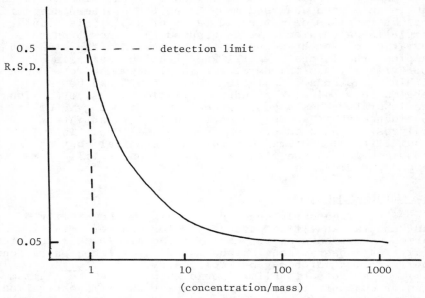

Figure 2. Typical variation of relative standard deviation of measurement vs. concentration or mass of analyte.

14.7 Preconcentration and Separation

In most applications of instrumentation to the identification and quantification of chemical species, whether this be in the laboratory or in an industrial environment, sampling and sample preparation for analysis frequently poses problems. These problems may be associated with difficulties in obtaining a representative sample or with the presence in the material to be examined of components which would give rise to non-specific interference in the analytical technique to be employed. A common procedure which is required under such circumstances is the physical separation of the component of the matrix to be examined prior to its examination by an instrumental technique of analysis. While this separation is achieved it is also frequently convenient to effect a pre-concentration of the component so that its determination by the instrumental technique chosen is more readily effected.

Great strides have been made in the development of highly selective analytical measurement techniques. However the analytical scientist is called upon increasingly to deal with more and more complex samples; as a result separation steps can be necessary even with highly selective instrumental methods such as neutron activation analysis or atomic absorption spectroscopy. As mentioned above, separation of a component of interest can also be used for its concentration in the medium to be examined by the instrumental technique of concern; this effectively increases the sensitivity of the analytical technique employed. Although separation procedures are often not, strictly speaking, instrumental techniques, they frequently comprise an integral part of an instrumental procedure and in fact may be incorporated into an instrumental design. The physical separation of components in a matrix to be examined by an analytical instrumentation technique is usually accomplished (a) by precipitation, either of the component to be determined or of the bulk matrix, (b) solvent extraction, (c) chromatography (gas, liquid or solid) separation of components physically and/or with respect to time and (d) electrophoresis. These commonly employed techniques of physical separation are used prior to many on-line, instrumental and automated techniques of determination of particular inorganic and organic components. The physico-chemical principles of these separations are well understood and the types of procedure developed are found in many applications from laboratory analysis to on-line control and even in assistance in production and process engineering.

14.8 Modern Techniques of Analytical Instrumentation

The past 20 years has seen a rapid growth of sophisticated techniques of instrumental analysis depending on a variety of physico-chemical principles for their operation, sensitivity and selectivity. Many of these techniques enable the determination with high precision of very small concentrations or masses of material. Thus, not infrequently techniques are encountered which can provide for the determination of parts per billion (nanograms/millilitre) of materials as diverse as steroids, toxic gases and trace metallic elements. Many of the procedures employed are found in routine operation in quality control and other control instrumentation systems at present employed industrially. A broad classification of the most common types of technique which have been developed for these purposes may be:

(a)   Selective spectroscopic techniques
(b)   Selective electrochemical techniques
(c)   Selective chromatographic techniques

The above types of technique are frequently automated for routine analytical process and quality control. The principal types of automated analytical system which are at present employed are concerned with (1) batch analysis, usually off-line, and (2) continuous analysis of flow streams. Continuous analyser systems take two forms depending on whether the flow is segmented or unsegmented. Numerous systems are available for these purposes and will be reviewed in my later lecture. This brief overview has attempted to demonstrate the principles underlying the requirements for chemically selective measurements for analysis; the fundamental principles are common with those involved in any process or engineering measurement system but the method by which chemical selectivity is attained may be quite different and will be dealt with subsequently.

Chapter 15

Analytical instruments

Professor G. F. Kirkbright

15.1 INTRODUCTION

As mentioned earlier a very wide range of physico-chemical phenomena are exploited for transuction and measurement of chemical parameters in modern analytical instrumentation and for generation of "Analytical" signals which can be made the basis of reliable measurements. These effects are invariably spectrochemical, electrochemical or based on differential absorption or partition coefficients for chemical components (chromatography). It is clearly impossible in this type of review to detail the principles and applications of all the techniques which might be encountered in major industrial laboratories and on-line in the process industries. An attempt will be made here, however, to describe some of the principles upon which these techniques operate and to illustrate some common features in instrumentation employed in many of the systems which have been developed.

15.2 Selective Spectroscopic Techniques

The word spectroscopy is widely used to mean the separation, detection and recording of energy changes (resonance peaks) involving nuclei, atoms or molecules. These changes are due to the emission, absorption or scattering of electromagnetic radiation or particles. Spectrometry is thus that branch of physical science which treats the measurement of spectra. The experimental applications of spectroscopic methods in chemical measurement are diverse but all have in common the interaction of electromagnetic radiation with the quantised energy states of matter. Thus we may wish to determine the molecular structure of a material, make an elemental analysis or verify the presence of certain features in the structure. Additionally quantitation may be required to determine the amount of material present in a given matrix which has a particular chemical nature. With the diversity of possible applications the chemist/engineer must choose a particular spectroscopic method pertinent to the problem at hand; his knowledge of the possible energy states of matter in particular configurations and the particular wavelengths of electromagnetic radiation that interact with these states is then of paramount importance.

The theoretical basis for the interaction between radiation and the energy states of matter is the quantised nature of energy transfer from the radiation field to matter and vice versa. Matter, composed of "particles" like protons, neutrons and electrons, sometimes behaves like a wave; radiation, a self propagating wave of crossed electric and magnetic fields, sometimes behaves like a particle. This seeming paradox is reconciled in the quantum theory which is used to calculate quantised energy states.

The wave-like character of radiation can be described by its wavelength, λ: by the wave number $\bar{\nu}$, which represents the number of waves per unit of distance (the reciprocal of wavelength); by the speed at which the wave front advances, the velocity V; and by the number of waves passing a given point in unit time, the frequency, ν. The relationship amongst these properties is given by

$$\bar{\nu} = \frac{1}{\lambda} = \frac{\nu}{V}$$

The velocity of electromagnetic waves in a vacuum is c (the speed of light, which is about 3×10^{10} cm.seconds^{-1}); the velocity in any other medium is lower. The absorption or emission of electromagnetic energy occurs when an atomic system changes from one energy state to another. The absorption or emission process corresponds to a photon of radiant energy, $h\nu = E-E'$ where $E-E'$ is the difference in energy between two states of the system. Similar processes occur for molecular species superimposed on which is the complexity of simultaneous changes in vibrational and rotational energy levels. Figure 1 represents the electromagnetic spectrum from D.C. to the X-ray region; the frequency and wavelength ranges corresponding to those used by different analytical spectroscopic methods are also shown. Table 1 illustrates for the principal spectroscopic methods employed in analytical instrumentation the corresponding energy states of matter or the basis of the phenomenon observed.

Figure 2 shows a schematic block diagram of a basic spectrometer; this includes some typical devices and conditions employed. Thus most spectroscopic systems used in analytical instrumentation conform to the general arrangement of source/sample/analyte/detector/signal display. The practical realisation of such spectrometers is sometimes complex (as evidenced from Figure 3 which shows a schematic diagram of a typical double-beam spectrometer used for infra-red analytical spectroscopy).

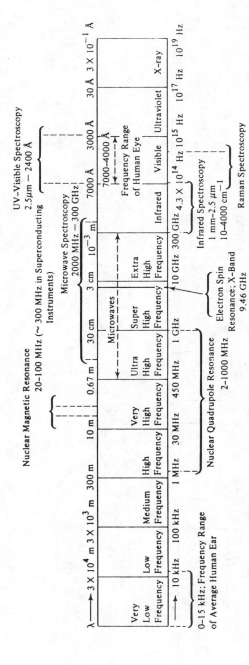

Figure 1. The analytically useful sections of the electromagnetic spectrum (from 'Instrumental Analysis, H.H.Bauer, G.D.Christian and James E.O'Reilly, Allyn and Bacon Inc., U.S.A., 1978).

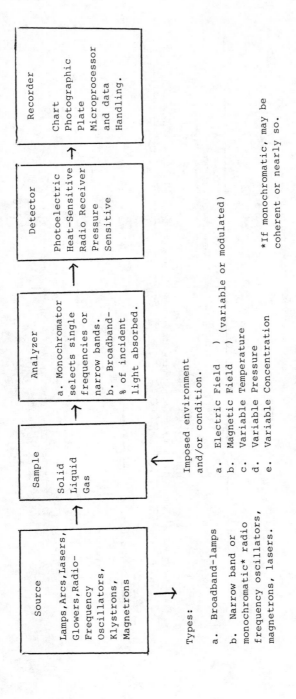

Figure 2. Schematic representing a block diagram of most spectrometers including some typical devices and conditions.

Analytical instruments 217

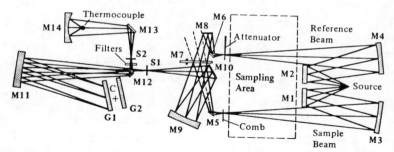

The symbols M1, M2... indicate mirrors; S1 and S2 indicate slits; and G1 and G2 indicate gratings.

Figure 3. Schematic diagram of a typical double-beam infrared spectrometer (Perkin-Elmer Corp., Norwalk, U.S.A.).

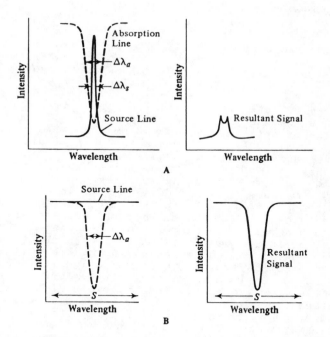

Figure 4. Principle of operation of atomic absorption spectrometer: A, line source and B, continuum source.

Atomic absorption (A) a sharp-line source and (B) a spectral-continuum source. $\Delta\lambda_a$ = absorption line half-width; $\Delta\lambda_s$ = source line half-width; S = spectral bandwidth of monochromator.

TABLE 1   Spectroscopic Methods and Corresponding Energy States of Matter or Basis of Phenomenon Observed

Method	Basis
Nuclear Magnetic Resonance	Nuclear Spin Coupling with applied magnetic field.
Microwave Spectroscopy	Rotation of Molecules.
Electron Spin Resonance	Spin Coupling of unpaired electrons with applied magnetic field.
Infra Red and Raman Spectroscopy	Rotation and Vibration of Molecules, Electronic Transitions for large molecules.
Ultra Violet-Visible Spectroscopy/fluorescence	Electronic transitions in atoms and molecules.
X-ray and electron spectroscopy	Electronic transitions Diffraction, reflection, fluorescence of x-ray radiation from atoms.
Mass spectrometry	Separation by mass/charge ratio of ion fragments produced from atoms and molecules.

Analytical instrumentation for atomic spectroscopy was first devised in the early nineteenth Century - flame spectroscopic techniques were responsible for the discovery of numerous elements in the periodic table. Routine analytical procedures, particularly in the metallurgical and mineral industries have been based on emission spectrometry using spectrograph equipment for almost 100 years. Recent developments in emission spectroscopy are typified by the application of very high temperature plasma sources, in particular the inductively-coupled plasma. Some examples of the sensitivity and utility of this type of system will be presented. Perhaps, however, the technique which has brought highly selective trace element analysis into common usage in industry, medicine and environmental analysis has been atomic absorption spectroscopy. The principle of operation of this technique with flame or furnace atom cells for sample introduction is illustrated in Figure 4 and a representation of an atomic absorption spectrometer is shown in Figure 5.

Analytical instruments 219

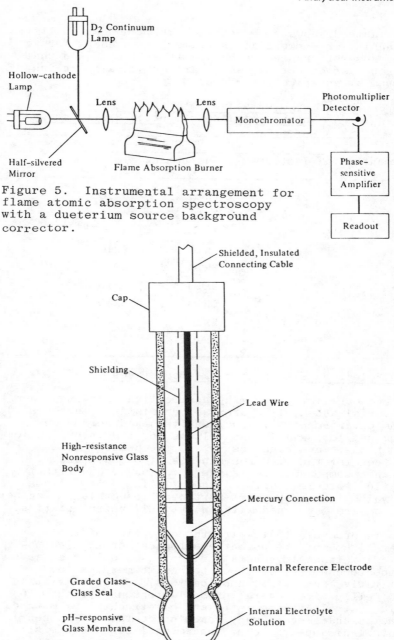

Figure 5. Instrumental arrangement for flame atomic absorption spectroscopy with a dueterium source background corrector.

Figure 6. Constructional details of a glass pH electrode.

Table 2 shows some representative detection limits obtainable for commercial instrumentation using furnace atomic absorption spectroscopy. It will be noted that the absolute mass detection limits are principally in the picogram or sub-picogram range.

Table 2   Representative Detection Limits for Furnace Atomic-Absorption Spectroscopy

Element	Detection Limit, g		
	Instrument A	Instrument B	Instrument C
Ag	2×10^{-13}	2×10^{-13}	5×10^{-13}
Al	3×10^{-12}	5×10^{-12}	4×10^{-13}
Cd	1×10^{-13}	1×10^{-13}	4×10^{-13}
Cr	5×10^{-12}	2×10^{-12}	4×10^{-13}
Cu	1×10^{-12}	3×10^{-13}	8×10^{-13}
Mn	1×10^{-12}	2×10^{-13}	4×10^{-13}
Ni	1×10^{-11}	5×10^{-12}	2.5×10^{-12}
Tl	1×10^{-11}	2×10^{-12}	-
Zn	6×10^{-14}	1×10^{-13}	-

15.3 Selective Electrochemical Techniques

Electrochemistry has many applications and uses in both fundamental and applied areas of chemistry - in the study of the environment, clinical analysis and corrosion phenomena, for example, as well as for the investigation of mechanisms and kinetics of electrochemical reactions, as a tool for the electrosynthesis of organic and inorganic compounds and in the solution of quantitative analytical problems both on and off-line. It is this last area which is perhaps most pertinent to consideration here where we are concerned with analytical instrumentation.

It is probable that, with the exception of the nearly universal use of the potentiometric pH meter, electrochemical methods in general are not as widely used as are spectrochemical or chromatographic methods for quantitative analytical applications. Bauer, Christian and O'Reilly cite several probable reasons for this (6). One is that electrochemistry and electrochemical methods are not emphasised in many teaching courses. The nearly universal disappearance of fundamental electrochemistry from elementary physical chemistry courses has been cited; in contrast to this, the interaction of electromagnetic radiation with matter is usually covered in many elementary courses. Electrochemical theory is really no more complex than, but probably not so well unified as at present, spectrochemical theory. A second reason given may be that spectrochemical methods appear somewhat more amenable to automation than electrochemical methods (6). An extreme example of this can be seen in the clinical analysis laboratory. In 1971, for instance, it has been cited that one particular hospital performed nearly half a million chemical tests, 91% of which were done with spectrochemical methods and instrumentation. This, of course, was due to its use of automated clinical analysers which depend primarily on optical measurements. There are, however, many places in the process industries where electrochemical techniques can provide essentially the same information as other techniques, thus offering an alternative approach, and other occasions when the best answer to the problem in hand can only be provided by electrochemistry.

Elemental electrochemical analysis is generally specific for a particular chemical form of an element. For example with a mixture of iron (II) and iron (III), electrochemical analysis can reveal the amount of each form present where most elemental spectrochemical or radiochemical methods simply give the total amount of iron present almost irrespective of its chemical form. Depending on the analytical problem at hand or the question to be answered, one particular method may be better. Many examples of this may be quoted; for example, mercury is a serious environmental pollutant. Elemental or inorganic forms of mercury represent toxic hazards but organic mercury compounds (such as methyl mercury or dimethyl mercury) give rise to more serious hazards. Electrochemical techniques frequently offer the ability to differentiate between different bound forms of the elements. Another advantage of many electrochemical methods is that they respond to the activity of a chemical species rather than to its concentration. An example where this may be of importance is the calcium level in blood serum. Ion-selective electrodes respond to free, aquated calcium ions. The usual clinical method for serum calcium is flame photometry, which measures the total calcium present, including a large amount present as protein-bound calcium. The more important physiological parameter, the measure of the effective level of calcium actually available for participation in various enzymatic reactions, may be the free calcium ion level. In another example, lead is cumulatively a toxic substance; plants grown in lead-laden soils can accumulate high levels of lead. If these plants are then eaten by humans, toxic levels of lead may be reached. Lead in heavy clay soils is, however, much less available for absorption by plants than is lead in a more sandy soil. Perhaps the more useful measure of the agricultural value of trace-metal contaminated land is the metal-ion activity rather than the total metal concentration. It can safely be said that in recent years there has been a renaissance of interest in quantitative electrochemical techniques and instrumentation for these. This has been brought about primarily by two factors: (a) the development of ion-selective potentiometric electrodes which can monitor quantitatively most of the common ionic species in solution, and (b) the introduction of a new generation of inexpensive commercial voltammetric instrumentation based on pulse methods. This latter instrumentation has increased the sensitivity of electrochemical techniques by several orders of magnitude.

15.3.1 Classification of Electrochemical Methods.
Electrochemical methods can be divided into two classes: those involving no net current flow ("potentiometric") and all others. In potentiometry the equilibrium thermodynamic potential of a system is measured without causing electrolysis or current drain on the system - otherwise the existing equilibrium would be disturbed. In all other methods a voltage or current is applied to an electrode and the resultant current flow through, or voltage change of, the system is monitored. The applied waveform is often quite complex. Although this approach may be more complicated than is the case in potentiometry, there are advantages in that it is not necessary to deal with the particular equilibrium characteristics of the system. By forcing the system to respond electrochemically to an external stimulus considerable analytical control over the system is obtained.

15.3.2 Potentiometry. Potentiometry - in which the electric potentials of electrochemical cells are measured - represents one of the oldest methods of chemical analysis still in wide use. The early analytical applications of potentiometry were essentially those used to detect end points in volumetric analysis. More extensive use of direct potentiometric methods came after the development of the glass electrode for pH measurements in 1909. In recent years, several new classes of ion-selective sensors have been introduced, beginning with glass electrodes more or less selectively responsive to other univalent cations (for example, sodium and ammonium). More recently solid-state crystalline electrodes for ions such as fluoride, silver and sulphide, and liquid ion-exchange membrane electrodes responsive to many simple and complex ions (for example, calcium and perchlorate) provide electro-chemical probes responsive to a wide variety of ionic species. Figure 6 shows a schematic representation of the construction of a typical glass pH electrode and Figures 7 and 8 are representations of the construction of a crystal-sensor ion-selective electrode and a liquid-liquid ion exchange electrode respectively. Table 3, for example, illustrates some typical properties of commercially available crystalline solid-state electrodes.

15.3.3 Polarography and Voltammetry. As pointed out above, in potentiometric techniques employed in analytical instrumentation no net current flow in the system is

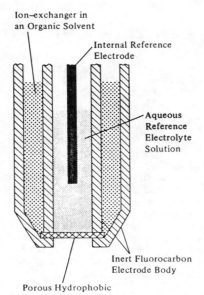

Figure 7. Schematic representation of the construction of a liquid-liquid ion-exchange electrode.

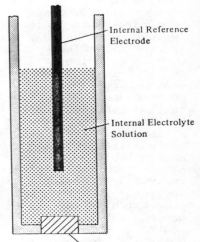

Figure 8. Schematic representation of the construction of a crystal-sensor ion-selective electrode.

Electrode	Concentration Range (M)	Activity Limit $C > 10^{-6} M$	Interferences
F^-	10^0–10^{-6}		$OH^- < 0.1 \, F^-$
Ag^+ or S^{2-}	10^0–10^{-7}	10^{-20}	$Hg^{2+} < 10^{-7} M$
Cl^-	10^0–5×10^{-5}	10^{-6}	$S^{2-} < 10^{-7} M$; trace Br^-, I^-, CN^- permitted
Br^-	10^0–5×10^{-6}	10^{-7}	$S^{2-} < 10^{-7} M$; $I^- < 2 \times 10^{-4} \, Br^-$
I^-	10^0–2×10^{-7}	10^{-10}	$S^{2-} < 10^{-7} M$
CN^-	10^{-2}–10^{-6}		$S^{2-} < 10^{-7} M$; $I^- < 0.1 \, CN^-$; $Br^- < 5 \times 10^3 \, CN^-$
SCN^-	10^0–5×10^{-6}		$I^-, S^{2-} < 10^{-7} M$; $Br^- < 3 \times 10^{-3} \, SCN^-$; $CN^-, S_2O_3^{2-} < 10^{-2} \, SCN^-$; $NH_3 < 0.1 \, SCN^-$; $OH^- < SCN^-$; $Cl^- < 20 \, SCN^-$
Cd^{2+}	10^0–10^{-7}	10^{-10}	$Ag^+, Hg^{2+}, Cu^{2+} < 10^{-7} M$
Cu^{2+}	10^0–10^{-8}	10^{-10}	$S^{2-}, Ag^+, Hg^{2+} < 10^{-7} M$
Pb^{2+}	10^0–10^{-7}	10^{-10}	$Ag^+, Hg^{2+}, Cu^{2+} < 10^{-7} M$

Table 3. Performance Characteristics of Commercial Crystalline Solid State Electrodes.

permitted or observed. In voltammetry, current/voltage curves are recorded when a gradually changing voltage is applied to an electrochemical cell containing the solution to be examined, a stable reference electrode and a small-area working or indicator electrode. Usually the voltage is increased linearly with time; such curves are frequently called voltammograms. In the special case where the indicator electrode takes the form of a dropping mercury electrode (DME) the technique is known as polarography and the current/voltage curves are called polarograms. With the advent of low cost commercial instrumentation and the introduction of some modern variants of the polarographic method - pulse polarography, stripping analysis etc. - the use of voltammetric methods for quantitative analytical measurements has increased considerably. The newer variations of the method can permit selective analysis at the parts per billion (ng.ml^{-1}) level for a variety of organic and inorganic species. The areas of application include environmental and toxicological studies, biochemistry and pharmaceutical process work, geology and routine industrial quality control. Figure 9 shows an illustration of an early polarographic instrument of classical form utilising a dropping mercury electrode; Figure 10 is a photograph of a modern voltammetric electrochemical system. Figure 11 shows a typical polarogram obtained from the classical form of instrumentation and the manner in which it might be used to monitor the nature of the species and its concentration in the solution under examination. Figure 12 shows the manner in which modern systems employ pulse and differential pulse techniques to generate the analytical signals obtained at the working electrode. Such modern microscopic controlled instrumentation has enormously improved the routine applicability of the technique of polarography.

15.4 Selective Chromatographic Techniques

We employ the term "chromatography" to describe those methods by which two or more compounds in a mixture physically separate themselves by distribution between two phases; (a) a stationary phase, which can be a solid or a liquid supported on a solid, and (b) a mobile phase, either a gas or a liquid, which flows continuously around the stationary phase. The separation of individual components in this manner results primarily from differences in their affinity for the stationary phase.

In liquid chromatography (LC) the flowing or mobile phase is a liquid, whereas in gas chromatography (GC) it is a gas.

15.4.1 Liquid Chromatography. There are several types of liquid chromatography, each distinguished by the predominant mechanism involved in the process of

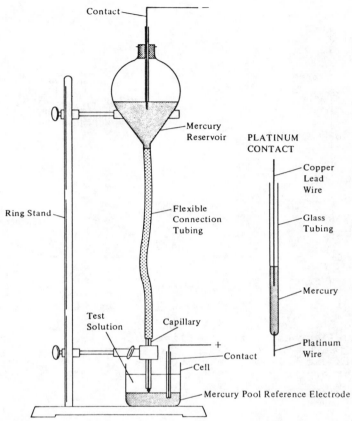

Figure 9. Classical experimental arrangement for DC polarography using a dropping mercury electrode (DME).

Figure 10. Modern microprocessor-controlled polarograph.

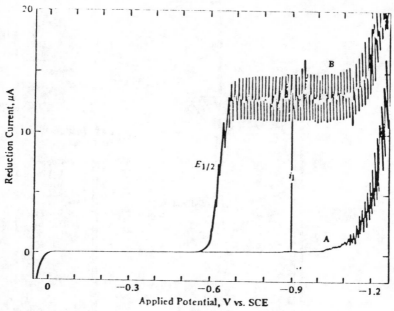

Figure 11. Typical polarogram. Curve A: Background, residual current, or supporting electrolyte curve 1M HCl). Curve B: Polarogram of 0.5mM Cd^{2+} in 1 M HCl. $E_{1/2}$ is the half-wave potential and i_l is the limiting current of the polarogram.

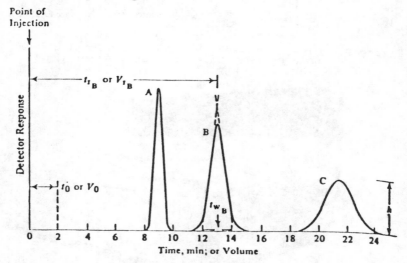

Figure 13. Chromatogram of a three-component mixture. t_0 = time for solvent to traverse the column, t_{IB} = retention time of substance B, t_{wB} = peak basewidth of substance B, h = peak height. Units can also be given in terms of volume rather than time: V_O, V_{IB}, V_{wB}, and so forth.

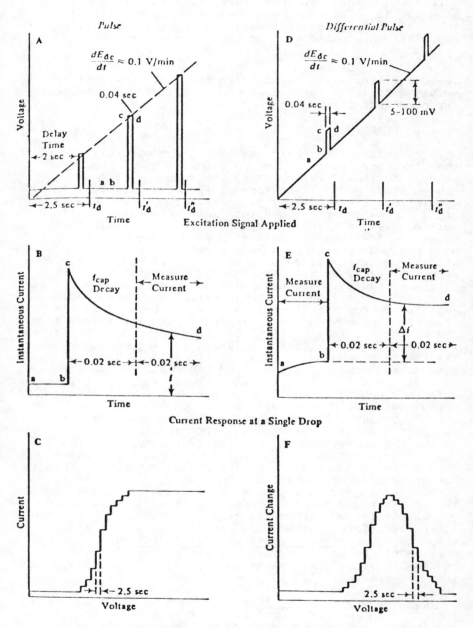

Figure 12. Current Voltage Curves obtained in pulse and differential pulse polarography.

separation. The stationary phase governs the separation mode as the solute molecule usually has some affinity for the stationary phase it transfers from the mobile phase to the stationary phase setting up the equilibrium

$$X_m \rightleftharpoons X_s$$

These equilibria are established for each component in the solute material. The corresponding distribution coefficient for component X is given by

$$K_x = \{X\}_s/\{X\}_m$$

A large value of K_x indicates that the component favours the stationary phase and moves slowly through the column whereas for small values of K_x the component favours the mobile phase and is therefore eluted more rapidly. In <u>adsorption chromatography</u> often referred to as liquid-solid chromatography the solute interacts with the fixed active sites on a solid adsorbent used as the stationary phase the adsorbent is packed in a column or may be spread on a plate or impregnated into a porous paper. The active sites, e.g. the surface silanol groups of silica gel, generally interact with the polar functional groups of the solute compounds. The non-polar portion of a molecule has only a minor influence on the separation. Thus liquid-solid chromatography is well suited for separating general classes of compounds (for example, separating amines from hydrocarbons). In <u>partition chromatography</u>, also referred to as liquid-liquid chromatography, the solute molecules distribute themselves between two immiscible liquid phases, the stationary phase and the mobile phase, according to their relative solubilities. The stationary phase is uniformly spread on an inert support - a porous or non-porous non particulate solid or porous paper (paper chromatography). To avoid mixing of the two phases the two partitioning liquids must differ greatly in polarity. If the stationary liquid is polar (for example, decanol) and the mobile phase is non-polar (for example, hexane) then polar components are retained more strongly; this is the usual mode of operation. On the other hand, if the stationary liquid is non;polar (for example, decane) and the mobile phase polar (for example, water), polar components favour the mobile phase and elute faster. The latter technique (which has a reversed polarity) is referred to as <u>reverse phase liquid-liquid chromatography</u>. Because of the subtle effects of solubility differences liquid-liquid chromatography is well suited for separating isomeric species.

<u>Ion-Exchange Chromatography</u> is based on the affinity of ions in solution for oppositely charged ions on the stationary phase. Ion-exchange packings consist of a porous solid phase, usually a resin, onto which ionic groups are chemically bonded. The mobile phase is usually a buffered aqueous solution containing a counter-ion whose charge is opposite to that of the surface groups - i.e. it

has the same charge as the solute - but which is in charge
equilibrium with the resin in the form of an ion-pair.
Competition between the solute and the counter ion for the
ionic site governs chromatographic retention. Ion-exchange
chromatography has found wide application for the
separation of metallic ions and in biological systems for
separating water-soluble ionic compounds such as proteins,
amino acids and common anions. A typical liquid
chromatogram for a 3-component mixture is depicted in
Figure 13. This illustrates some of the terms used in
chromatographic practice.

15.4.2 Gas Chromatography.

Perhaps more than any other
technique of instrumental analysis gas chromatography has
proved to be of critical importance in many industries and
disciplines. Gas chromatography is a type of partition
chromatography, similar in many ways to other techniques of
this kind such as liquid chromatography, paper chromatography, etc. The distinguishing features of gas
chromatography are that the mobile phase is a gas and that
the motion of the component bands, in the direction of
"chromatographic development", involves the forced
diffusion of the respective substances in their vapour
phases. Many of the differences between for example liquid
chromatography and gas chromatography are due to the
physical properties of the mobile phase - for instance its
viscosity, acidity and compressibility. The basis for
differential zone-migration remains the same: two
components will migrate at different rates in the same
chromatographic system if their distribution constants are
different. Figure 14 shows a block diagram schematic of a
dual-column gas chromatograph showing the essential parts.
A large number of methods of detection of the eluted
components from gas chromatography columns exist. The
thermal conductivity (hot wire) detector was one of the
earlier forms employed although in widespread use is also
the flame ionisation detector. A very sensitive and
selective detector is the electron capture detector. These
three types of detector are illustrated in Figure 15. As
an example of performance for modern gas chromatography
Figure 16 shows chromatograms obtained for a standard
mixture of polycyclic aromatic hydrocarbons and of a coke-
furnace emission.

Automation of Analytical Instrumentation.

Automated
instruments are generally classed as <u>continuous</u> or
<u>discrete</u> (batch) depending on the nature of their
operation. A continuous instrument senses some physical
or chemical property by directly observing the sample
yielding an output that is a smooth function of time. A
discrete instrument works upon a batch-loaded sample and
supplies information only after each batch. Each derives
its operating principles from conventional analytical
procedures and must include provision for continuous
unattended operation: receiving samples, performing
selective chemical analyses under uncontrolled

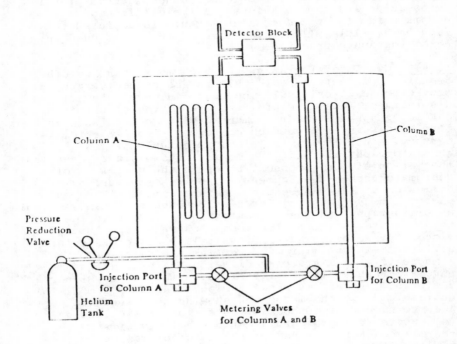

Figure 14. Block diagram of a dual-column gas chromatograph showing essential parts.

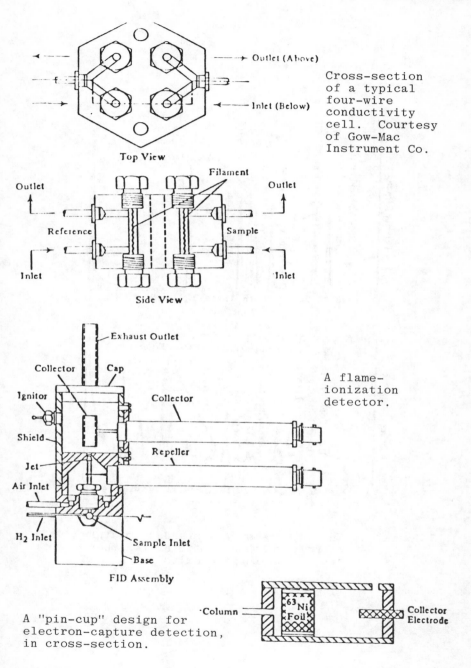

Figure 15. Some configurations of detector assembly for gas chromatography.

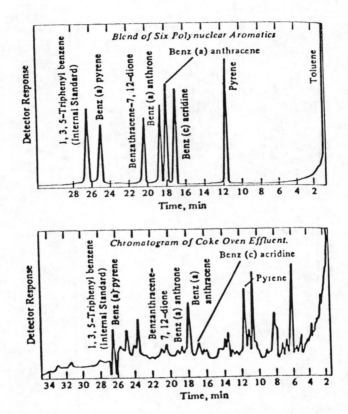

Figure 16. Chromatograms of a standard mixture of polycyclic aromatic hydrocarbons and a coke-furnace emission.

enviromental conditions and communicating with monitoring or control equipment.

A clear distinction should be made between <u>automatic</u> and <u>automated</u> devices. Automatic devices cause required acts to be performed at given points in the operation without human intervention. For instance, an <u>automatic titrator</u> records a titration curve or simply stops a titration at an end point by mechanical or electrical means (such as a relay) instead of manually. Automated devices, on the other hand, replace human manipulative effort by mechanical and instrumental devices regulated by feedback of information; thus the apparatus is self-monitoring or self-balancing. An <u>automated titrator</u> may be intended to maintain a sample at some preselected (set point) state, for example pH = 8. To do this the pH of the solution is sensed and compared to a set point of pH 8 and acid or base is added continuously to keep the sample pH at the set point.

In the past automated instruments were not well accepted because of their limited capability and reliability. However, because of the increased complexity and number of clinical, industrial and other types of samples requiring analysis, classical (n-automated) techniques, as well as automated techniques, have been improved in capability. Well established instruments such as infrared analysers, gas chromatographs, ion-selective electrode systems and automatic wet-chemical analysers can now measure quite complex species and mixtures. Reliability has also increased because the maturity of solid-state electronics has brought easier data handling and equipment maintenance.

The chemical instrumentation discussed in the preceding sections can all be utilised in automated systems. The choice is largely dictated by economics and the applicability of an instrument to the proposed problem. It would be difficult to review all of the instrumental systems used in automated control in the space available. Spectroscopic electrochemical and chromatographic systems are the most widely employed and a review of the principles of operation of these has been presented here in small compass. I have drawn here extensively on several standard texts concerned with instrumental analysis - notably 'Instrumental Analysis' by Bauer, Christian and O'Reilley (6). I gratefully acknowledge their permission to quote from this work and to reproduce diagrams from this text.

Selected References for Further Reading

1. K.J.Clevett, Hand Book of Process Stream Analysis, New York, Halstead Press, 1973.

2. R.Hicks, J.R.Schenkin and M.Steinrauf, Laboratory Instrumentation,New York,Harper and Row, 1974.

3. T.S.Light,Industrial Analysis and Control with Ion-Selective Electrodes in "Ion-Selective Electrodes", National Bureau for Standards Special Publication 314, Washington D.C., 1969.

4. L.W.Lee, Elementary Principles of Laboratory Instruments, Third Edition, St.Louis,Missouri, C.V.Mosby, 1974.

5. F.G.Shinskey, pH and pIon Control in Process and Waste Streams, John Wiley and Sons, New York, 1973.

6. H.H.Bauer, G.D.Christian and James O'Reilly, 'Instrumental Analysis', Allyn and Bacon Inc., 1978.

Chapter 16
Microprocessors in instrumentation
Dr. R. M. Henry

16.1 INTRODUCTION

Not so very long ago instrumentation was all about transducers. It was essentially a branch of applied physics. Since that time before microprocessors, the whole subject has expanded to embrace new areas, new techniques and specialists from other fields. The transducers are every bit as important as they ever were, but now other areas have grown in importance.

Even today instrumentation is still largely about direct measurement. However, with the sort of online signal processing and computation available through microprocessors we can expand instrumentation to include a whole new field, inferential measurement.

Inferential measurement means combining the output of two or more transducers in some way; it may be necessary to study the outputs over some period of time. You have already met one example of inferential measurement in the correlation flow meter. Despite its name, it does not measure flow. It estimates the transit time between the noise signals from two transducers mounted on the pipe. Knowing the separation, it is a simple matter to infer the velocity of flow. Typically 5 s of data is needed to provide a repeatable result. Before microprocessors this was an expensive measurement. Now it can be carried out at a modest cost.

The subject of instrumentation has been expanded in other ways as well. Data presentation used to be a simple choice between a meter and a chart recorder. Now it is bedevilled with ergonomic considerations and man-machine interactions. Besides meters and chart recorders there are now digital displays, LED strips, and the TV screen. The latter can present text and graphics in colour or monochrome. One page of display can be rapidly replaced with another and another. Information comes to the operator. Gone are the days of walking round enormous control rooms with a clip-board.

In principle, there is nothing that we can do with a microprocessor that we could not have done ten years ago with a minicomputer. It is the price of the hardware that has changed making the computing power so cheap that we cannot afford not to use it.

The advances of large scale integration that gave us the microprocessor also make it possible to produce low cost dedicated digital circuits. This is obviously only for high volume production and a good example would be time measurement. The electronic approach using quartz crystal drives has now totally displaced the mechanical timekeeper developed over more than 300 years. Cash registers have suffered a similar demise.

This shows how digital electronics can create totally new solutions to problems. This approach can also generate new products and, in particular, new measurements. For example, the automatic single lens reflex camera uses electronics in the light measurement and in the control of shutter speed. By linking one function to the other the exposure can be made automatic. It is a staggering thought that today, this relatively new product is the largest volume measurement made by microelectronics. This is another direction in which instrumentation has expanded.

Expansion has been most dramatic in totally new areas where measurements were being made for the first time. In the process industries, where instrumentation really got started some 50 years ago, the take up of the new technology has been rather slower to get going. There are good reasons for this conservatism. The whole running of the plant depends on the reliability of the instrumentation. Everyone would rather someone else was the first to use the latest instrument on plant. The pattern is rather similar to, the uptake of a new computer control scheme. A newcomer has a long slow start before gaining general acceptance. This is because many in the industry prefer the old proven system on the basis of "Better the devil you know ...."

Another way in which instrumentation has expanded is due entirely to microprocessors. Once the decision is made to incorporate a processor, then it is so easy to add all sorts of extras such as self-testing, engineer's diagnostics etc. Soon the instrument stops being a simple measuring device and becomes an instrumentation system.

Enough of generalising: it is not a good way of getting these ideas across. I shall now outline a number of applications from which the reader must reach his own conclusions, bearing these thoughts in mind.

16.2. MONITORING PATIENTS RECOVERING FROM CARDIAC SURGERY

Perhaps medicine offers the best examples of the difference between what one wants to know and what one can measure. This is certainly true of the present example.

Patients recovering from cardiac surgery do so in an intensive care centre. There they are connected to a monitor which measures:-
1) Mean blood pressure
2) Mean heart rate
3) Core temperature
4) Peripheral temperature
5) Urine output

All these measurements are know to be relevant to the patient's state of recovery. Another way of phrasing this is to say that the information on the patient's recovery is distributed between the measurements. There may however be duplication and delay between measurements and this is what makes appraisal of the measurements a tricky business. The medics learn from long experience and from each other. Such learning is locked up in their minds and is not available to program a microprocessor. The transfer of such knowledge to create intelligent systems is an important research area promising great returns.

In this case regression analysis was applied to many, many hours of patients' records. This showed that all the information was contained within the first two measurements and that these were the first to respond to a change in the condition of the patient.

However, these particular measurements suffer from artefacts, sudden short-lived, random, large swings of no meaning whatsoever. It is these artefacts which make patient monitoring so difficult. Each channel of the monitoring equipment already had pre-settable alarm limits. If these are set too close there will be numerous false alarms and the operator will simply turn down the volume and ignore future alarms. If the alarm limits are set wider then most alarms will be genuine, but they will only occur after a substantial deterioration in the patient's condition.

What is needed is a program which will look at both the MHR and the MBP together. Any large change affecting only one channel and lasting less than 6 seconds is assumed to be an artefact and is ignored. Much more statistical design is required before the algorithm is ready to monitor. This includes an on-line learning phase in which adaptation is made to the individual patient.

The software for these algorithms has been implemented on a Z-80 microprocessor and the system has been connected to the existing monitor to take over the alarm monitoring function. The initial appraisal, after several months on-line, is most encouraging. (Lu 1983)

This example is in fact an inferential measurement. The information comes from two transducers and is spread over a period of time. The computer initially adjusts itself to the patterns of the individual patient during a learning phase. Thereafter it monitors the patient's recovery never failing to detect a deterioration, often before the nursing staff, and very rarely alarming when there is no real cause for concern. This is an example of an intelligent knowledge based system, an area currently being supported by the government's Information Technology program.

16.3. ANALYSIS OF PROCESS STREAM COMPOSITION

This is a complicated subject deserving a whole book rather than a subsection of a single chapter.

Two methods are available: gas chromatography and mass spectrometry. The former is the method most used; it costs around #20,000 per measurement and has been developed over the last twenty years. It is now a proven and reliable measurement.

Mass spectrometry has been around for a similar length of time but until quite recently was seen as a laboratory instrument. It is more expensive than the gas chromatograph, typically around #40,000, but offers advantages of speed and flexibility. It relies heavily on micro-computing and looks set to make in-roads into the market for on-line analysers, especially once its reliability is demonstrated.

These two techniques are reviewed together because one shows the application of a microcomputer to an existing product whilst the other shows how a processor is conceived as an integral part of an analysis system. First though, a brief introduction to the two techniques.

16.3.1 Gas Chromatography

A tube contains an inert material the surface of which is covered with the stationary phase, a material chosen so that different components of the gas will 'stick' for different periods. The whole apparatus is run at some constant temperature, often elevated to ensure that the input stream for analysis is gaseous.

The column is constantly purged with an inert carrier gas. At the start of analysis a measured quantity of the sample is introduced. The different components proceed along the column at different rates which are known for a given stationary phase. At the exit the separated components are detected by one of a number of possible techniques. This measurement is recorded on a chart and the area under each peak corresponds to the quantity of each component.

Clearly this is a case for integration and the first attempts used analogue integrators. High quality integrators were needed to prevent drift during the period of the analysis, a matter of 2 to 5 minutes in most cases.

The obvious place to apply microprocessors lies in the integration. Going digital totally circumvents the drift problem and saves money at the same time. Now that the decision has been made the next question is "What else can we make the processor do." Every manufacturer has his own answer to that question. Selling to an industry which wants improved instruments and proven reliability, both at the same time, most manufacturers have settled for a 'bolt-on' approach. All the sampling, temperature control, column, carrier gas equipment etc. is unaltered (giving proven reliability) but the analysis and presentation of the results is improved with graphical displays, automatic calibration, digital integration etc.

16.3.2 Mass Spectrometry

Mass spectrometry works be examining the movement of charged particles in a magnetic field. If an element is being analysed then all the ions have ths same mass:charge ratio and all follow the same path in the magnetic field. When a molecule is ionised by a stream of high energy electrons it breaks into a number of charged fragments of different mass.

PERSONNEL & ENVIRONMENTAL TRace-gas Analyser

PETRA DATA

PETRA DATA is the software written for the microcomputer driven PETRA. It enables a number of programmes to be used for analysis, calibration, VDU and hard copy print-out in various formats. The software design is such that it can be operated by a person without computer expertise and simple single letter commands can give access to any programme listed.

Programmes are loaded into the microcomputer's 48K memory by the use of a compact floppy disk drive unit. Many output facilities are available including visual display unit, video copier, printer, printer plotter and disk storage.

Illustrated are examples of just some of the programmes available to the user.

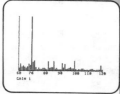

Spectrum Histogram

The spectrum histogram routine provides a mass spectrum upto 120 amu wide in an easy-to-read histogram form. The spectrum is continually updated whilst the display is being viewed. This programme aids the identification of products that may be present other than those known by the user.

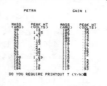

Table Display

A numerical measurement of the intensities of each peak in the spectrum histogram can be obtained by using the table display command. The table presents a list of mass peaks measured together with their intensities.

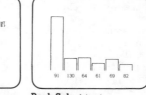

Peak Select (run)

A histogram output of selected peaks is displayed when this programme is used. Various dwell times can be used to optimise the time of analysis and the accuracy of each measurement.

Peak Select (table)

When a limited number of compounds in the atmosphere need to be analysed, up to 15 characteristic peaks can be selected in order to acquire a faster analysis than that obtained by the spectrum histogram programme. A conversion from the signal intensity of each characteristic peak to parts per million can automatically be carried out by introducing a known level of calibrant sample into the calibration loop.

Data Output

The Occupational Hygiene chemist generally requires information in the form of time weighted averages (TWA), standard deviation (σ) and short-term exposure levels (STEL) during a pre-determined period.

Calibration Calculation

A simple programme for calculating the quantity of 'calibrant' sample required can be used if the calibrants molecular weight and density are known.

Fig. 16.1 EXAMPLES OF GRAPHICAL AND NUMERICAL DISPLAYS
Reproduced by kind permission of V G Gas Analysis Ltd.

For example methanol, CH_3OH (mass 32) breaks into CH_3^+ (mass 15), $CH2OH^+$ (mass 31) and CHO^+ (mass 29). The proportions are predictable and so the spectrum (intensity vs mass number) can be used to recognise the initial product.

All this has to be carried out at high vacuum to prevent reactions taking place between the ions. To measure the intensity at a single mass number takes only 1 second. The gas chromatograph is two orders of magnitude slower. The analysis of mixtures presents some interesting problems for now there are superimposed different amounts of two or more spectra. As the number of components increases so the regression becomes lengthier. This clearly calls for some on-line 'number-crunching' to deal with the general case. Happily, in many cases it may be possible to obtain the result by looking at mass numbers unique to given components.

As with gas chromatography there is a need to display information in both graphical and numerical form. Figure 16.1 shows some of the displays available on VG Gas Analysis Ltd's PETRA system for the analysis of trace gases in the environment.

With mass spectrometry there is a greater computing need than with gas chromatography. This shows up in the closer integration of the microprocessor in the design of the system.

Comparisons should surely be quantitative, in terms of function and performance, but are we always that objective. Do we ever modify the objective function to favour the system we prefer for subjective reasons. I confess that I prefer well-integrated microprocessor applications rather than the 'stuck on as an after-thought' type. This is purely subjective but then is there anyone who can ever be truly objective.

16.4.   MEASUREMENT ON A PNEUMATIC CONVEYING RIG

Much of the work described here for use on pneumatic conveyors has been developed on a pneumatic conveying test plant at Bradford University. This plant has been designed specifically for computer monitoring and control so it would be right to start with a description of this plant. Later some uses to which this plant has been put will be described followed by details of a Z-80 based correlator contrasting its speed and flexibility against larger, more expensive machines.

16.4.1   Overall Plant Description

Figure 16.2 shows a schematic diagram of the plant. Basically the system is a negative pressure pneumatic conveyor with air drawn through it by a compressor. The pipework, in 75mm I.D. aluminium, is assembled in short sections to allow great flexibility of the physical layout. The simplest form is a vertical C shape. Solids enter the system with the air at one end of the C. At the other end the air and solids are separated using a cyclone and the solids are fed via a rotary valve into a hopper for recirculation. Thus continuous operation can be achieved.

The solids and air feedrates can be separately controlled, this is described later in the paper. Apart from temperature, pressure and weight transducers the plant is equipped with electrodynamic transducers to sense the motion of the solids directly. This is described later as well.

Microprocessors in instrumentation 241

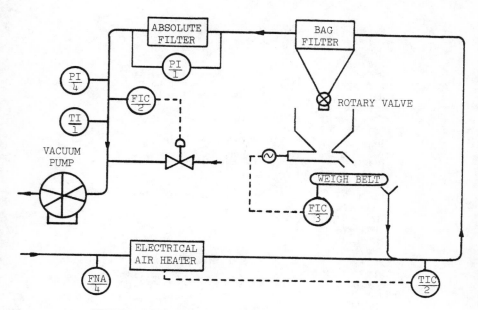

Fig. 16.2 GENERAL LAYOUT OF EXPERIMENTAL PLANT

16.4.2 The Air Feed System

The air feed system consists of a Hick Hargreaves sliding vane compressor drawing air through the plant pipework, the cyclone, a filter, an orifice plate and then past an air bleed valve into the compressor inlet. The bleed valve allows air to be sucked directly into the compressor instead of it having to go through the rest of the plant pipework. In this way the air feedrate throught he plant can be adjusted.

The orifice plate is fitted with transducers to measure absolute upstream presure, differential pressure and downstream temperature. The differential pressure transducer output is used as the input signal of a conventional controller and all three signals are fed as inputs to the computer. The control signal is fed, via an I to P converter, to a pneumatic actuator on the bleed valve. Using the conventional controller a square root factor is involved due to the non-linearity of the orifice plate. However, the computer can use all three signals to calculate exactly the air mass flowrate.

16.4.3 The Orifice Plate Calculations

The standard text for orifice plate design and operation is British Standard BS1042. Therein scaling and correction factors are included for accurate calculation of air mass flowrate or velocity. Figure 16.3 lists a section of a program in BASIC used by the computer to read the transducer signals and calculate the air mass flowrate.

```
1000 REM                subroutine to measure the mass flow of air
1010 REM                and return an answer in W in kg/s
1020 C=.6082
1030 Z=1.005
1040 REM                D0 is the diameter of the orifice.
1050 D0=2.75
1060 REM                D1 is the diameter of the pipe.
1070 D1=5
1080 REM                M is the diameter ratio.
1090 M=(D0/D1)^2
1100 REM                Y is the ratio of specific heats.
1110 Y=1.4
1120 REM                E is the velocity of approach factor.
1130 E=1/SQR(1-M*M)
1140 P9=0
1150 REM                Get the upstream pressure P1 from a/d ch.11.
1160 FOR I=1 TO 20
1170 AIN(,P1,1,,11,1)
1180 P9=P9+P1
1190 NEXT I
1200 P1=P9/20
1210 REM                Scale the voltage to give a pressure in P.S.I.
1220 P2=(P1-.25)*14.6959
1230 REM                Get the differential pressure H from a/d ch.9.
1240 H9=0
1250 FOR I=1 TO 20
1260 AIN(,H,1,,9,1)
1270 H9=H9+H
1280 NEXT I
1290 H=H9/20
1300 REM                Scale to inches water.
1310 H1=(H-.25)*20
1320 REM                Scale to P.S.I.
1330 H2=H1*.0361273
1340 REM                Calculate the pressure ratio h("H2O)/P(psi).
1350 R=H1/P2
1360 REM                Get the temperature from a/d ch. 8.
1370 T9=0
1380 FOR I=1 TO 20
1390 AIN(,T,1,,8,1)
1400 T9=T9+T
1410 NEXT I
1420 T=T9/20
1430 REM                Scale the temp. to degrees Kelvin.
1440 T=(T-.25)*100+273
1450 REM                Perform a density correction on the air.
1460 REM                Density of air is 0.08072 lb/ft3 at STP
1470 D=.08072*(P1-.25)*(273/T)
1480 REM                Calculate the expansivity of the air from figures
1490 REM                taken from the chart on pg 122 of the standard.
1500 E1=1-(R/.775)*.0103
1510 REM                The formula following uses:
1520 REM                        pressure in     p.s.i           P2
1530 REM                        diff press in   " h2o           H1
1540 REM                        diameters in    inches          D0
1550 REM                        density in      lb/ft3          D
1560 W=359.2*C*Z*E*E1*D0*D0*SQR(H1*D)
1570 W=W*1.26000E-04
```

Fig. 16.3 Calculation of air mass flow using BS 1042

16.5. CORRELATION FLOW MEASUREMENT

Correlation flow measurement depends on measuring the transit time of natural irregularities in the flow. This makes it a useful method for many of those difficult applications where conventional techniques reach their limit. One such example would be two-phase flow.

Two transducers are used and they can be anything which responds to natural noise. Electrical conductivity, capacitance and static charge have all been used sucessfully. Signal processing removes all the dc component leaving a noise signal with zero mean value.

The transducers need to be spaced quite close together (typically no more than 1 pipe diameter) since the noise pattern is continually evolving. Resolution demands measuring the transit time with a quantisation step of about 1%. This in turn suggests a sampling rate an order of magnitude higher than the bandwidth of the noise signals dictates.

We need a microprocessor not only to correlate, but to deal with the problem of 'too much data and not much information'. Without attention to this latter problem one would need something far more powerful than even the current 16-bit microprocessors.

Hardware correlation could be used but at some stage one would still need a microprocessor even if only to present the results and convert from transit time to velocity. With cheap hardware it is sometimes cheaper to be computationally inefficient. Nevertheless, the elegant solution has a certain attraction (like a flame to a moth, perhaps!). Here dedicated hardware would be the cheapest solution provided volume was sufficient to justify chip design. So far, it is not and a dedicated software approach proved the least costly. There are two ways of proceeding and both will be outlined.

The first simplification involves a reduction in the sample resolution. The ultimate reduction is a single bit denoting the polarity of the signal. The correlation still works albeit with a poorer signal:noise ratio and is known as polarity correlation. This is used for the first method which introduces a novel idea. Instead of (over)sampling a clock is read whenever a zero-crossing occurs. Now, time resolution depends on the clock speed whilst the number of zero-crossings depends on the signal bandwidth. This decoupling achieves the required reduction of data without information loss and requires only a special correlation algorithm designed to work directly from zero-crossing time data.

The second method requires two different sampling rates, one eight times faster than the other. The upstream transducer is sampled slowly and the downstream one quickly. This again achieves resolution whilst the data is reduced by a factor of eight. This algorithm can handle either polarity data or 4-bit quantised data. Multiplication of 4-bits x 4-bits is achieved using a 256 entry look-up table, the data being used to address the table.

In both cases the processors support graphic display facilities in addition to their numerical function.

16.5.1 Zero-crossing time algorithm for polarity correlation

To recap, data comes from two quite closely spaced transducers designed to measure random variations in the flow pattern. The object is to estimate the transit time between the two transducers. Sampling at the Nyquist frequency would give very poor time resolution and so higher sampling rates are used, commonly 10 to 20 times higher. This, of course provides lots more data but no more information. If logged, the polarity data would be in the form of long strings of 0's and 1's. This is not very convenient to handle and a more suitable representation would be to record the transition times. There is no space saving for each transition or zero-crossing requires 8 or 16 bits. The big saving comes because there are far fewer items to handle. Note also that the number of items does not change as the clock rate is changed. There is a decoupling which we are exploiting.

The zero-crossing time algorithm works directly from data in the form of zero-crossing times. From this data it directly calculates the nett agreement between the signals. See figure 16.4. A special interface is used to log the data.

One very interesting feature of polarity correlation is that the nett agreement can be divided by the total observation time to give the normalised polarity correlation coefficient. This figure is valuable because it is a measure of the reliability of the correlation measurement.

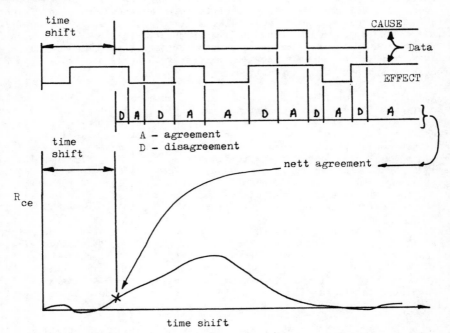

Fig. 16.4 Calculation of correlation coefficients from zero-crossing times

16.5.2 Implementation of the algorithm

A minimum configuration would consist of data capture hardware and software, the correlation algorithm and some method of displaying the result. To cover a range of applications some flexibility is needed in the sampling rate (actually a clockrate because there is no sampling) and the amount of data logged. These two items are controlled from thumbwheel switches.

This then represents the very minimum possible configuration for the correlator. In this form the device would be very limited and there are so many nice features we can add.

One serious limitation of the basic form just described is that it does not give the normalised correlation coefficient that goes with the peak position. There is a need to present more than one item of information and this leads to the use of a TV screen for display/communication.

16.5.2.1 Presentation of data

The microprocessor chosen for this application was a 4 MHz Z-80a and implementation was done using NASCOM 2 boards. We wished to use a standard proven piece of hardware.

One feature of the NASCOM 2 system is that a 1K block of memory is mapped directly to a TV monitor screen giving 16 lines each of 48 characters. Figure 16.5 shows just how much can be written on a screen at once. Add to this that any number of other frames can be presented at no extra hardware cost and the logic of using a TV screen becomes irrefutable.

```
  012345678901234567890123456789012345678901234567
1 THIS SHOWS A SCREENFUL OF DATA. THE FRAME IS  1
2 48 CHARACTERS BY 16 LINES. NOT ONLY CAN THE   2
3 SCREEN CARRY TEXT, IT CAN ALSO BE USED FOR    3
4 GRAPHICS. TWO TYPES ARE AVAILABLE; THE FIRST  4
5 USES ORDINARY CHARACTERS AND HAS POOR RESOL-  5
6 UTION WHILST THE SECOND DEPENDS ON THE USE    6
7 OF A GRAPHICS CHARACTER GENERATOR. THIS       7
8 GIVES FAR BETTER RESOLUTION                   8
9                                               9
0                                               0
1                                               1
2                                               2
3                                               3
4                                               4
5 12345678901234567890123456789012345678901234567
```

Figure 16.5 16 lines, 48 character display

16.5.2.2 Functions, facilities and extras

Having taken the decision to use a TV screen then one might as well use it to the full. Numbers can be presented with titles and units. More importantly, some elementary graphics can be included.

Other features include more functions than just correlation, peak search procedure, peak tracking and a back record display, a sort of chart recorder mode. These are all outlined below.

16.5.2.3 Extra functions Besides straight cross correlation the correlator also provides auto-correlation, a peak search procedure which speeds up computation by a factor of about 7 times, a back record which presents previous peak positions in the form of a bar chart, a self test feature and finally a histogram of the distribution of zero crossing intervals. This latter function provides a rapid check on incoming data and immediately shows up any transducer faults such a mis-matching of the channels.

Altogether there are eight functions and they can run either in 'single experiment' mode or continuous mode. In the former the processor halts after each correlation; with the latter data collection and analysis loop endlessly.

The eight functions are:-

1. Peak search presenting peak graphically and numerically
2. Peak search presenting back record
3. Peak search presenting histogram of upstream data
4. Peak search presenting histogram of downstream data
5. Correlation - everypoint, display stops at peak
6. Self test
7. Auto-correlation of upstream data
8. Auto-correlation of downstream data

16.5.2.4 Graphics These are quite rudimentary but well worth having. Figure 16.6 shows a correlation peak. The peak position is the principal feature and by allowing the eye to judge the shape of the peak any fault is immediately apparent.

An auto-scaling routine is used and it is important to refer to the table in the bottom right hand corner of the screen which shows the correlation coefficient.

```
              THE CORRELOGRAM
     01A2 49      !--------
     0290 50      !-----------
     03AA 51      !----------------
     04E4 52      !---------------------
     0638 53      !-------------------------
     06FC 54      !-----------------------------
     0758 55      !--------------------------------
     075C 56      !--------------------------------
     071C 57      !-------------------------------
     06D0 58      !-----------------------------
     0644 59      !--------------------------
     05BE 60      !----------PEAK POSITION=0056
     051C 61      !----------CLOCK PERIOD =0250 us
     0416 62      !----------PEAK VALUE   =01884
     02E6 63      !----------MAX.COFF.Rxy =0.28
```

Fig. 16.6 Correlation function - every point

16.5.2.5 Peak search

The data is 'over sampled' as already explained. This means that the peak will be quite wide. The peak position can be found faster using a search procedure which has two stages. First the whole field is scanned and one point at least will lie on the side of the peak. Then points around this first maximum are scanned using a bisection type of search. The result is shown in figure 16.7. Using this procedure computation time (including display management) is about 0.4 s compared with 3.5 s for every point.

```
                THE CORRELOGRAM
       0000  47    !
       0168  48    !------
       0000  49    !
       0000  50    !
       0000  51    !
       05A6  52    !----------------------
       06D4  53    !--------------------------
       0772  54    !----------------------------
       079C  55    !-----------------------------
       0724  56    !---------------------------
       0000  57    !
       05E6  58    !---------PEAK POSITION=0055
       0000  59    !           CLOCK PERIOD =0250 us
       0432  60    !---------PEAK VALUE     =01948
       0000  61    !           MAX.COFF.Rxy =0.28
```

Fig. 16.7 Correlation function - peak search

16.5.2.6 Data checking and result checking

Data is checked quickly and simply by generatinmg a histogram of zero crossing intervals (obtained by differencing zero crossing times). A typical result is shown in figure 16.8 which represents band limited Gaussian noise. Both channels should have similar histograms and any departure from this would indicate a transducer fault or transducer mis-matching.

```
                    HISTOGRAM OF CAUSE
                      10        20        30        40
                 ---------!---------!---------!---------!  (frequency)
             04  DDDDDDDDDDDDDDDD
   class     08  DDDDDDDDDDDDDDDDDDDDDDDDDDDDDDDDDDDDD
   interval 12  DDDDDDDDDDDDDDDDD
   (clock   16  DDDDDD
   periods) 20  DD
             24  D
             28
             32
             36
             40
             44
             48                     NO.OF POINTS=0255
             52                     SCALING :X3
```

Fig. 16.8 Histogram of zerocrossing times

Besides checking input data the results are also checked. This is done using the normalised correlation coefficient. The result will be accepted if the coefficient lies above an upper threshold and rejected if it lies below a lower threshold. Results with coefficients between the two thresholds will only be accepted if they show close agreement with previously accepted values.

The results can also be checked in relation to one another by inspection of the back record. This is shown in figure 16.9. Poor repeatability results from poor quality data; by increasing the observation time and improvement in the correlation signal:noise ratio can be obtained and this may overcome the problem.

```
           BACK RECORD
    56 !-----------------------------------
    56 !-----------------------------------
    55 !-----------------------------------
    56 !-----------------------------------
    55 !-----------------------------------
    55 !-----------------------------------
    56 !-----------------------------------
    55 !-----------------------------------
    56 !-----------------------------------
    55 !-----------------------------------
    55 !-----------------------------------
    56 !-----------------------------------
    57 !-----------------------------------
    55 !-----------------------------------
    55 !-----------------------------------
```

Fig. 16.9   Backrecord of latest 15 peak positions

16.5.2.7 <u>Further result processing</u> The processor applies an exponential filter to the peak positions in order to obtain a single averaged value from which the velocity is computed. This value is available as a voltage to drive a meter and as a 4 - 20 mA current loop output.

16.5.2.8 <u>Peak tracking</u> Peak tracking correlators do not enjoy a very high reputation. There are problems centring on what happens when a peak is lost and how it can be recovered.

When we set out this correlator was not intended to have peak tracking option. However, the idea of changing the clock rate (equivalent to a sampling rate) to keep the peak position in the mid-range was attractive. It was easy enough to try and worked very well. Unlike other attempts at peak tracking there is practically no chance of losing the peak because each evaluation scans the whole range.

This feature is available as an optional extra.

16.5.3 Features for engineers

The previous section described the facilities available when the correlator is in use. There are a number of other important features which were built into the software package.

16.5.3.1 Options and epROM firing
It was not very long before our 'customers' in the university were asking for slight modifications to the software. For example, one might want a different set of clock rates and another a different pair of threshold values. Such modifications did not involve any new code; they simply involved changing constants in the assembly code program.

By itself this was not much of a problem and we did our best to oblige. Pretty soon we found we had lots of different epROM sets and it was quite difficult to tell which was which.

The answer was to write a small program which would interrogate the appropriate memory locations and present a summary of the options on the screen. See figure 16.10. This small program was fired into some space left on the last epROM and made life much easier. There was no extra cost.

We still had the problem of firing customised epROMs and that was solved using a BASIC program on the development system. Rather than simply copy the software from disc to RAM and then fire, the procedure was as follows:-

1. Copy correlation software from disc to RAM
2. Load BASIC
3. Run BASIC program. This program conducta a question and answer session modifying the correlation software in RAM as indicated by the answers.
4. Fire epROMs

We can now provide a wide range of options with very little effort and any chip set can be interrogated to tell what options have been specified.

```
            DIRECTORY
THE -8-CLOCK RATES ARE:
2857 us  2000 us  1428 us  1000 us  Clock periods in microseconds
0667 us  0500 us  0333 us  0250 us

   IZ   OZ   L   RT  VM RXY% E%  NP     IZ, OZ inner and outer zones
   40   70  20   98  10  05  04  10 18 80 0112  for peak tracking
                                         L Transducer spacing (cms)
                                         RT reject time for data (secs)
                                         VM maximum velocity (m/s)
                                         RXY% confidence limits
                                         E% exponential averaging factor
                                         NP no. of points correlated
```

Fig. 16.10 Summary of correlator software options as displayed on screen

16.5.3.2 Diagnostics When the system was being developed a number of test routines were written. These tested the counter timer circuit, the thumb-wheel switches on the front panel and logged data (without subsequent analysis). When we had finished our development work there was a little bit of space on the epROMs and so these were included. The start addresses of these programs appear with the option information.

For the future I would expect most systems to come with their own diagnostics. There is no extra software cost because the routines were already written and the extra #2 for another epROM (if necessary) would seem like money well spent.

16.5.3.3 Driving software from BASIC At some stage the software features have to be frozen and development work stopped. Despite all one's efforts to meet all reasonably foreseeable demands it will not be long before someone comes along and says "If only I could do ....". How nice to be ablke to say "You can" and send him packing!

We solved this problem by making the whold package callable from the Nascom's Microsoft BASIC using the USR routine. It was one of the best things we did and has allowed much other work to be easily implemented.

When calling the package the user specifies (by a single argument) whether or not the package graphics are to be supressed. On return this argument holds the peak position. Any other information can be obtained using PEEK and DEEK. A full memory map is available. Using these techniques we have carried out work on pattern recognition of repetitive waveforms and some other work on non-linear identification.

The usual criticism that BASIC is slow and inefficient is hardly valid when the processor is spending most of its time running the highly efficient assembly code of the correlation package. When used this way Basic is being used to improve the user interface and provide axtra facilities that are not time critical.

16.5.3.4 Communication The NASCOM 2 has a serial port which is not used by the correlation package. It is available for any communication required with other processors. Programming might be in BASIC or a combination of BASIC and assembler.

16.6 ACKNOWLEDGMENTS

My acknowledgments to Dr S M Said who spent many hours on the software described above as part of his doctoral program. Also to SERC for grant number GR/A 7451.7 which supported this work and to the Military Technical College, Baghdad, Iraq who supported Dr Said during his research.

REFERENCES

1. Henry R M, Al Chalabi L A M "Microprocessor Applications to velocity measurement by cross-correlation" Acta IMEKO 1979 Moscow p 88

2. Lu R "A microprocessor based immediate alarm system for post cardiac surgical patients" PhD thesis University of Bradford Nov 1983

Chapter 17
A case study the development of an instrument to measure coal seam thickness

V. M. Thomas

17.1 INTRODUCTION

This paper is a largely chronological account of the development over a 20 year period of one of the key measurement techniques required by the mining industry in it's advance towards the automation of coal face operations. The basic requirement is to determine the position of the cutting drum of a mining machine in relation to the upper and lower seam boundaries, ie the coal rock interfaces. The instruments described meet this requirement by providing a measure of the thickness of coal left between the upper seam boundary and the cut surface left by the machine as extraction proceeds.

Since one of the objects is to provide an illustration of a development to meet a real and difficult industrial measurement task, and to note the evolution with experience of the requirement as well as the solutions, it has been necessary to outline the mining context and in particular some of the features and benefits of automatic horizon control systems for coal cutting machines.

In a postscript to the main body of the paper, the experiences gained over this considerable period are used to provide a brief commentary upon the methodology of such research and development projects.

17.2 HORIZON CONTROL OF CUTTING MACHINES

17.2.1 The Longwall Mining System

The actual winning or cutting of coal in longwall mining occurs at the coalface, an excavation of approximately full seam thickness normally 1 to 2m in height and 100 to 300m in length. Figure 17.1 depicts such a coalface. A cutter-loader machine travels along the face depositing the mineral onto an armoured conveyor, essentially a series of segmented steel slide plates along which the coal is propelled by chain-driven transverse slats or flights. Access to the coalface for personnel, for transport of the mineral outbye (i.e. outwards from the workings to the shaft and the surface),

252 Case study

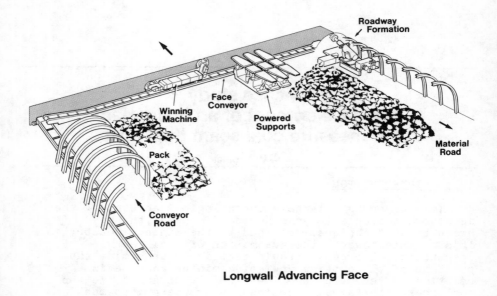

Fig. 17.1   Coal winning by the longwall method

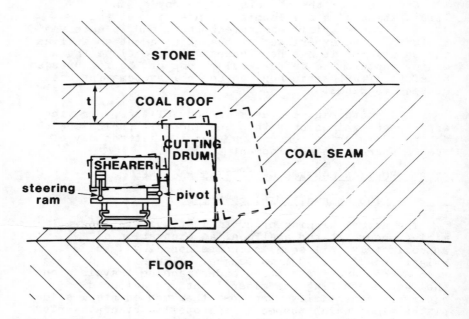

Fig. 17.2   Shearer extraction and steering action

transport for materials inbye, and for the circulation of ventilating air is provided by two 'gate' roadways, one at each end of the face.

Roof support at the coalface is limited to a strip a few metres in width, running the full length of the face. It is provided by a line of hydraulic chocks, each consisting of a number of hydraulically extendible legs which support the roof linked to the armoured face conveyor by an hydraulic jack.

Face operations consist of: (i) taking off a strip of coal from the exposed face some 0.5 to 1m wide, during a traverse of the shearer, a milling-type cutting machine; followed by (ii) advancing the conveyor a corresponding distance by exerting a forward thrust upon the conveyor using the hydraulic jacks attached to the chocks; and (iii) drawing up the chocks, one after the other, by lowering their supporting legs from the roof, applying hydraulic pressure to the ram linking them to the face conveyor to draw it forward, and subsequently reapplying pressure to the hydraulic legs to support the roof once again. After advancing the powered supports one after the other in this way, the unsupported roof soon collapses forming a rubble-filled waste area, or goaf, which subsequently slowly compacts.

17.2.2 The Need for Horizon Control

Horizon control of coal cutting machines has been a requirement ever since mechanised mining was introduced. In the early years of mechanisation, development was concentrated on the main components of the power-loader machines and face conveyors, ie the motors, gearboxes, cutting drums etc. Steering was a lower priority consideration and it was left to the machine operator to make the best of a difficult task. However, the attention of the Mining Research and Development Establishment (MRDE) was drawn to this problem as a key development in the attempt (with hindsight, rather a premature attempt in the mid 1960's) to advance towards the automation of coal face operations. This attempt became well-known in the mining industry under the title of ROLF, the Remotely Operated Longwall Face Project. These efforts were reported at a Symposium held by the Ass. of Mining Electrical and Mechanical Engineers (1), and the aspect of horizon control more specifically in the paper by Thomas and Pidgeon (2).

The need for horizon control is, at its simplest, that of cutting coal and not stone from a seam which may undulate, vary in thickness and not infrequently present other more troublesome impediments to mechanised mining such as faults, washouts or roof rolls. The mining of some stone arising, for example, from roadway drivage, or from cutting in the roof or floor of the seam, is an accepted part of coal mining; and coal preparation is often required to separate the coal from the reject dirt. However, the economic benefit of utilising the colliery transport, winding and preparation capacities to provide a higher percentage of saleable product is obvious.

But there are several less obvious benefits from good horizon control. When extraction remains within the seam dust is reduced, since the make of dust in cutting rock is much greater, and more damaging, than in cutting coal. Secondly, when smooth roof and floor profiles are left by a well-steered machine the continuing advance of coalface equipment as extraction proceeds is far kinder to the equipment, with benefits in lower maintenance and reduced delays. And, most importantly of all, the layer of coal normally required to be left at the roof in British mining practice can be minimised. This roof coal serves to assist in the critical process of controlling the break-up of the roof strata, which collapse behind the line of hydraulic face supports when the latter are advanced after a cut. Failure to control the roof in the immediate vicinity of machine, face conveyor etc, can result in rock falls and roof cavities with consequential delays and dangers. It is important for the mining engineer to be able to provide the roof coal thickness he needs; but equally important to avoid any excess since this reduces the proceeds potentially available. This is one of the prime benefits of good horizon control, and can be worth £0.5 m to £1 m per face annually.

17.2.3 Horizon Control of Cutting Machines

Figure 17.2 shows a cross-sectional view of the coalface and fixed drum shearer. This machine became the dominant power loading machine as mechanisation spread throughout the coalfields in the 1960's. (This was in turn displaced with the evolution of a further variant, the ranging drum shearer, in the 1970's.) As outlined above (section 17.2.1), extraction proceeds by the cutting of successive strips, the armoured face conveyor being thrust forward onto the cut floor preparatory to the next pass of the machine along the face. To provide a means of steering the fixed drum machine, the main body and the attached drum can be tilted with respect to an undercarriage which rides on the armoured conveyor track. Tilt is produced about a pivot, shown in Fig 17.2 by means of an hydraulic steering ram. An upward tilt of the drum results in a corresponding upward tilt of the conveyor when it is advanced onto that section of the floor; and similarly in reverse, except that loose cut coal can act as an infill for any undercut, considerably influencing the steering effect.

17.3 THE MEASUREMENT TASK

Sufficient has already been outlined to indicate that a measure is required of the roof coal thickness, t in

Fig. 17.2, suitable for use as an input to the horizon control system.

The requirements can be summarised as follows:

(i) thickness range — a measurement of t from 0 to at least 150 mm; and preferably to 400 mm or 500 mm if possible. (It is also valuable to measure depth of cut into the rock stratum if the machine cuts beyond the seam boundary i.e. to measure negative values of t, if possible)

(ii) speed of response — a time constant of measurement not exceeding 10 seconds (and preferably less) is required for machines which travel at speeds up to 6 m/min

(iii) mounting position — a transport lag is involved if the measurement is not made at the drum itself; the sensor needs to be mounted as near as practicable to the drum subject to an acceptable level of interference with the flow of coal from the drum

(iv) physical size — as small as practicable to ease the problem of the mounting position (iii), and to minimise projection beyond the machine profile since this increases the equipment vulnerability markedly

(v) electrical safety — the electrical circuits employed must be safe for operation in potentially explosive mine atmospheres. This is normally achieved by design and

(vi) working environment — certification to British Standards for intrinsically safe equipment (BS 1259, BS 4683, or BS 5501).

— the instrument must be robustly and appropriately constructed to operate reliably in an extremely adverse environment near a cutting drum; under water sprays, in a dusty atmosphere and subject to vibration (1g, 50 Hz to 5 KHz) and shock (40 g in 3 orthogonal planes)

17.4 POSSIBLE TECHNIQUES

By the mid 1960's the ROLF project was providing a specific stimulus to find solutions to the automatic horizon control problem.

The methods which were under consideration (even remote consideration) at this time for the measurement of roof coal thickness by the MRDE or others elsewhere were:

Technique	Measured variable
nucleonic	γ-ray backscatter flux from a radioactive source
sonic/ultrasonic	propagation time for wave reflections from the stone/coal interface at the seam boundary
electromagnetic	(i) apparent dielectric constant in a planar 2-material 'sandwich'
	(ii) propagation time for wave reflections from the coal/stone interface (radar)
mechanical	(i) penetration rate of a drill, or similar tool
	(ii) force on a cutting pick (not strictly a measure of coal thickness but a means of locating the

position of the cutting drum in the seam).

17.4.1 The γ-ray Backscatter Method

The density of coal (1300 to 1400 kg/m^3) is roughly half of the adjacent sedimentary mudstones or shales (2000 to 2400 kg/m^3) and this density contrast provides a good basis for a sensor based on γ-ray backscatter measurement. The method is one dependent upon 'bulk' properties, rather than interface detection; and preferable on that account, since carbonaceous shales (or dirty coal) can often provide a region of transition from coal to rock rather than a clean interface.

There was previous experience available to support further investigation and development of this technique. This related to early work on a γ-ray backscatter probe for use in boreholes to delineate coal seams; and more directly relevant, a floor coal thickness measurement using γ-ray backscatter. A schematic diagram of the experimental device produced by Becque and Thomas (3) for trials with a mole-like mining machine, the Collins Miner, is shown in Figure 17.3. The measurement characteristic and a general view of this Type 708 sensor are also shown.

Gamma radiation from a radioactive source, in this case 100 mCi of Americium-241, irradiated the coal and rock in its immediate vicinity. Back-scattered radiation was detected by a geiger tube, with a pulse-counting circuit to provide an indication of the radiation magnitude. Direct radiation from source to detector was minimised by a substantial intervening metallic shield. Operator protection from radiation during machine maintenance and similar non-operational periods was provided by an hydraulically operated rotary arming mechanism. In the 'safe' position the radioactive source was driven by a spring into a well screened interior location in the sensor, and only when machine hydraulic power was available was the sensor armed by rotating the source holder to the position behind a window in the sensor surface. From the characteristic response for this early Type 708 coal thickness probe it will be noted that the usable section of the curve extends only to about 70mm; the saturation value of about 100 counts/sec is a consequence of two principal factors important in the design of such sensors (i) the energy of the principal radiation (since penetration power increases with higher γ-energy values) and (ii) the geometry of the source, shield and detector arrangement - a larger source-detector separation increases the coal thickness value at saturation (though reducing the absolute count-rate value at saturation, thus forcing a compromise in practice).

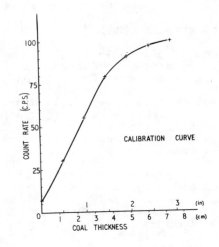

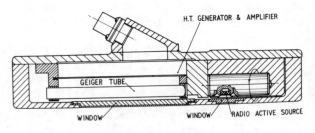

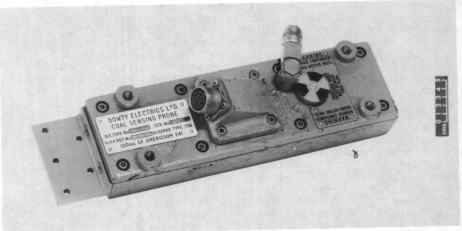

Fig. 17.3   Coal sensing probe Type 708

Against the requirement for a minimum thickness range of 150 mm further development was clearly necessary from the early Type 708 probe referred to above. However, with the prospect of meeting such a requirement and practical experience to back that judgement, this method was viewed as one worth pursuit.

17.4.2 Sonic/Ultrasonic Methods

We were aware of well-established non-destructive testing techniques based upon sonic or ultrasonic wave probing of materials such as castings, pressure vessel walls etc., indicating a possible method for coal thickness measurement. Again some background in coal was available. Extensive laboratory studies of ultrasonic transmission through coal samples had been undertaken some years before by Terry and Seaborne (4) as a means of determining the elastic moduli of coal. The outcome of these studies was a more realistic appreciation of one important limitation. This was the high attenuation and very noisy signals obtained with transmission of ultrasonic waves through coal in comparison with that through materials such as metal. The difference was ascribed to the small fissures in coal, due, presumably to its formation from compressed vegetable matter, which would give rise to a large number of randomly distributed scattering centres with the effective consequence noted - high attentuation and high noise. This was especially to be expected in the absence of the constraining overburden pressures, a condition which would apply at the exposed surfaces at the face.

Even more significant, efficient coupling from a transducer to launch waves into the coal would evidently be a major problem since the coal surface would be very rough and grooved by the picks; and some sort of sliding or rolling sensor would be needed for a moving machine. While a transducer could be envisaged using water as the coupling medium roughly sealed by some flexible skirt (cf hovercraft), this and the additional problem noted above did not make it an attractive approach.

17.4.3 Electromagnetic Methods

Two electromagnetic methods are in principle available, one based on a steady-state response to excitation, and the other on wave reflection techniques.

The steady state approach amounts to a method of measuring the dielectric and loss constants (complex permittivity) of a plane layer of one material (coal) resting on a half plane of a second (stone). The effective measurement depth will, of course, depend upon the geometry of the field excitation employed.

There was a significant added interest in this method since it was known that a thickness sensor based on this principle had been reported from Russia by Nosov (5); and a project was initiated to assess its prospects. The outcome was an indication that the moisture content of the mineral sample was a dominant factor - not surprisingly since water has a dielectric constant of 81 compared with 6 for dry stone and 4 for dry coal; and, as an added complication, the conductivity component of the complex permittivity varied with the ionic concentrations present. In spite of the expectation that the minerals would probably be saturated in their natural state, changes could be expected due to loss of moisture with time from exposure to the ventilation air-stream. A method relying upon the constant value of a dominant component, while measuring small changes in a subsidiary component did not appear an attractive candidate for practical development.

On the other hand, utilising a wave reflection method is difficult because of the extremely short travel times for electromagnetic wave propagation over distances of 100 mm to 500 mm, the two-way travel times being approx 0.7 to 3 nano sec. The technology to resolve such short travel times was by no means readily available in the 1960s and presents even today a formidable task. Considerable development of such a method occurred in the USA in the period 1975 - 80 and experimental sensors have been produced for evaluation as reported by Broussard and Schmidt (6). However, it was not a preferred method in the mid 1960s.

17.4.4 Mechanical Methods

At first sight, mechanical methods (i.e. other than the wave-based methods already considered in 17.4.2) seem impracticable. Detection of an underlying interface must seemingly be a remote-sensing problem. However, a possible basis for a mechanical solution was either proposed, or even tried, (the information was very sketchy) in the USA, arising probably from the experiences of practical mineworkers. A driller can often feel the difference between drilling in coal and in stone and possibly the rate of penetration or some such variable could 'capture' this difference. The idea is an intriguing one, and a 'low technology' solution could conceivably be developed; though with a number of problems to be resolved due to the machine movement (speeds up to 6 m/sec), and due to variations in drill bit wear and probably in the coal and stone characteristics as well. No resources were devoted even to assessing this possible method.

One other essentially mechanical technique was examined, and is still a live issue. It cannot strictly be classed as a coal thickness measure, but is relevant in providing the basic information required - drum position in a coal seam. The force on a pick in the cutting drum is the measured variable. This will vary

widely, due to i) the brittle fracture of a material such
as coal,   ii) to the banded characteristic of a coal seam
reflecting its natural geological origins, and, iii) to
the difference between coal and stone if the cut strays
up into roof or into the floor. In spite of the large
statistical variation because of (i) there is an
underlying consistent pattern of pick force against drum
rotational angle due to (ii) and (iii) which is a
'signature' for that seam, or at least for the seam in
that locality. In a fraction of seams there are very
clear features such as intergrown stone bands several
centimetres in thickness which create quite marked force
peaks as the pick traverses them. If the signature is
consistent enough over the panel of coal being extracted
i.e. over distances of several hundred metres, the
position of these peaks can be used to determine the
horizon of the cutting drum. Even in the absence of such
prominent bands, with adequate statistical processing,
the drum position can be derived.

In its more primitive marker-band form, this
technique was investigated in the late 1960s and shown to
be a practicable proposition. The proportion of seams
which displayed a sufficiently pronounced banded
structure was not determined however; its applicability
at the time seemed too restricted. However in the last 3
years, with the benefits of computer based processing of
the signals this situation has changed and the technique
has recently been pursued once again. There are two
important attractions in this technique. There is a
total absence of any transport lag - the measurement is
made at the drum itself; and secondly to the mining
engineer there are no interfering components (though to
the mechanical engineer, modifications are involved to
accommodate the force transducer and to provide for power
into, and signal transmission from the drum).

Summarising the position in the mid 1960's, the
preferred method of coal thickness measurement was the
nucleonic γ-ray backscatter technique. But from the
early experience of a floor probe in the Collins Miner
mole-type machine and some trials using this probe
mounted in a coal deflector plate towed behind the
cutting drum of a fixed drum shearer, futher development
was clearly called for. The aims were

(i)   to increase the coal thickness range to at least
      150 mm and preferably much more

(ii) to measure roof coal, not floor coal, since coal
      fines were a serious interfering factor with
      sensors riding on the floor.

17.5   TYPE 707 AND 709 γ-RADIATION BACKSCATTER PROBES

17.5.1   Type 707 Probe

The Type 707 γ-ray backscatter probe was the outcome

of this phase of development. The requirement to increase the maximum coal thickness which could be measured by a factor of 2 to 3 led to several changes from earlier sensors

(i) a higher γ-energy source was introduced to provide more penetrating radiation; instead of the previously used 100 mCi americium 241 source with 60 kev energy, a 5 mCi caesium-137 source with a main emission at 660 kev was used

(ii) the separation between source and detector was increased from 110 mm to 410 mm, unfortunately making the device longer and heavier

(iii) to compensate for the lower level of received radiation at the greater separation, a more sensitive detector than the Geiger tube was introduced, a scintillation crystal and photomultiplier.

A secondary consequence of the more powerful radioactive source was a much enlarged screening block to house this safely. Together with the accompanying general increase due to (ii) the mass was increased to a formidable 77 kg.

The main features of the design and the performance characteristic are shown in Figure 17.4.

It will be seen that the thickness range achieved at 150 mm was double that of the earlier probe (Type 708). Based on this increased performance, albeit in a considerably heavier and larger unit, prototype automatic steering systems were developed by Hartley (7) for fixed-drum shearer cutting machines in the late 1960s. The colliery trials which followed demonstrated the first successful automatic horizon control of a cutting machine anywhere in the world.

While the 707 probe development was a major element in the automatic horizon control project it must not be assumed that other aspects were minor or involved only routine design. The steering characteristics of such successive-strip type machines were much disputed and the underground confirmatory measurements were difficult to perform and labour intensive. Yet for sound control system understanding and design, they were essential. Another transducer was developed to measure the tilt of the cutting machine towards the coalface though this presented no comparable difficulty; and electrohydraulic machine controls and actuating rams were introduced as part of the overall control system.

The trials produced a number of alternative mounting positions and arrangements for the bulky coal thickness monitor. The fixed drum machine itself was used in

Case study 263

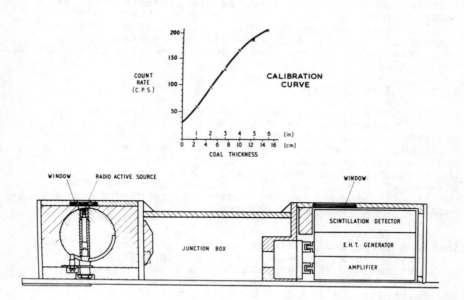

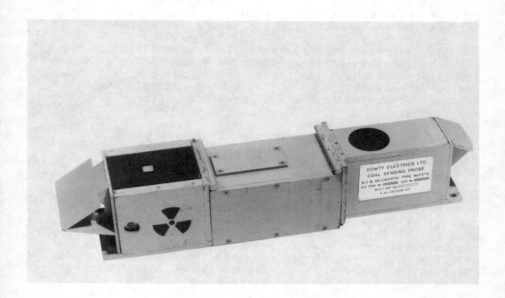

Fig. 17.4 Coal sensing probe Type 707

different ways and with different auxiliary equipment. For example some machines cut in one direction only, flitting back at higher speed to its starting position before commencing another cutting run. Others cut in each direction of travel. Each operational variant or sub-variant unfortunately made different sensor arrangements necessary. Two examples are illustrated in Figure 17.5.

17.5.2 Type 709 Probe

The technical success of the first shearer horizon control system also led to attempts to assess the applicability of the system within UK seams and mining practice. The outcome of these was a rough estimate that applicability could well be less than 25%, though some of the assumptions made were themselves in dispute. Both the operating experience, and the limited attempts at market assessment, however, confirmed the need for a still greater coal thickness range.

Responding to this need a second programme to increase range was initiated in 1969. The essential variable to raise the value of coal thickness at which backscatter detector saturation occurs was source-detector spacing. This was further increased to 580 mm making the finished overall probe with its protective end-fairings nearly 1 m in length. In addition to compensate for the lower absolute level of received radiation at the greater separation, a ten-fold increase of source strength, to 50 mCi, was necessary.

The performance is presented in Figure 17.6 together with that of the earlier probes for comparison. It shows the saturation thickness raised to 250 mm, giving a useful operational range of 170 mm, or possibly 200 mm. It should be noted that all of these characteristics assume that

(i) the coal is 'clean' i.e. free from bands of shale

(ii) the probe riding surface is in good contact with the coal surface.

The former of these is a natural consequence of the geological formation of the seam and must be accepted as part of the 'given conditions'. The latter is, on the contrary, within the designers hands to achieve. The severity of the effect depends upon the mean density and thickness of the dirt band, and on its position relative to the seam boundary. Deep kinks appear in the characteristic curve making automatic control at that thickness impractical.

The need to ensure good contact between the probe sensing surface and the coal is, in principle, straight

Fig. 17.5 Probe mounting arrangements

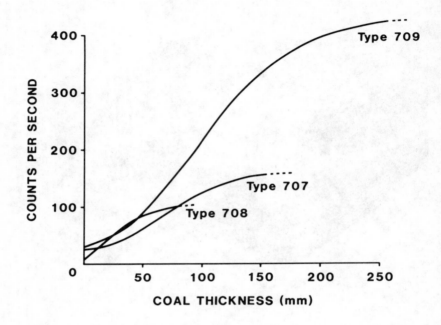

Fig 17.6   Probe characteristics

forward. However, bearing in mind the size and weight of the probe and the need to locate it near the cutting drum, the practical problems are formidable. As noted above, several different mounting positions were used and hydraulic jacking systems devised appropriately, to force the upper probe surface against the roof of the cut. The strength and ruggedness needed is evident from Figure 17.5; and the interference with normal machine operations at the face-ends due to such large components, protruding more than a metre beyond the end of the machine in some versions, can also be appreciated.

Notwithstanding these difficulties, between 1972 and 1982, production versions of these equipments were introduced and more than 100 face installations made, typically with 6 to 10 installations running at any one time. The reasons for this relatively small exploitation (1 to 2% of faces) were three-fold

   (a)   probe interference was a major limitation

   (b)   the number of suitable seams did not exceed 50% due to the roof-coal thickness required or other geological problems such as dirt bands

(c) the fixed drum machine was being steadily supplanted by a variant, the ranging drum shearer.

17.6. THE NATURAL GAMMA SENSOR

17.6.1 Conception

There occurred in 1969/70 a chance encounter which introduced an alternative proposal for a nucleonic method of coal thickness measurement. This occurred during a visit to a supplier and designer of probes which are used for logging exploratory boreholes in the search for coal and other minerals. Among the range of parameters measured by the probes was the natural gamma-radiation of the surrounding strata. The logs clearly delineated coal seams by a distinct lowering of the radiation level in contrast to that in the adjacent mudstones and shales. The level was very low, but clearly measurable, though of course ample integration time could be permitted in borehole logging, compared with the intended machine-based application.

A number of enquiries were conducted to establish the generality of these observations, and it was evident that with a few exceptions, notably for sandstone, roof and floor strata in the UK were shales with apparently similar levels of γ-activity, while coal had a low intrinsic activity. The concept of a sensor based on the natural radiation in the strata was thus born. The principle of the method is illustrated in the upper left section of Figure 17.7. Gamma radiation from the shale roof is attenuated in its pasage through the coal seam beneath; and a measure of the intervening coal thickness is provided by the magnitude of radiation at that position.

There were several important advantages that were in prospect from this new approach:-

- no radioactive source was required, other than that naturally occurring in the strata. This was beneficial in reducing the constraints and concerns which the earlier probes had caused.

- more, importantly, the radiation was received from a distributed planar source. The implications of this may not be immediately obvious but are vital to the subsequent success of the technique - the flux/field is invariant with distance from the plane surface (ignoring any attenuation effects, and considering only the geometric field factor). Analogous situations which may be more familiar are thermal radiation from a planar heat source, or electric field near a plane charged surface both of which are similarly independent of distance from the planar source (providing the source area is

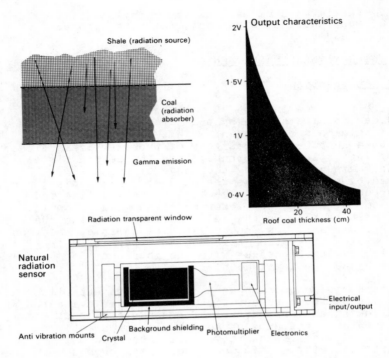

Fig. 17.7 Natural gamma coal thickness sensor

large enough). This offered the prospect of eliminating the need for the probe to be supported in close contact with the roof coal. Bearing in mind the severe application limitations incurred by the hydraulic elevating mechanisms and support arms needed for the Type 707 and 709 probes, this was a very attractive attribute of the natural gamma sensor.

There were also a number of problems to be solved before the new method could be applied. The main questions to settle were:-

- could a detector be devised, with an acceptable response time, for the very low level of radiation present?

- was the -activity constant enough over a panel of coal, since any significant local fluctuations especially along the coal face could make the technique unusable?

- was the variability of γ-activity from seam-to-seam small enough to make a single sensor design widely applicable?

17.6.2 Investigations and Outline Design of the Natural Gamma Probe

Apart from preliminary enquiries, it was not possible to devote adequate resources to resolving the main issues noted above until 1972. Detailed investigations into the technique followed during the years 1972-1978 and these have been reported by Wykes and Adsley (8).

The activity of the shale was shown to be due to three principal radio nuclides:- potassium, K^{40}, (typically 2.5%), thorium, $Th\ 232$ (10ppm); and uranium (radium 226 and daughters, 3ppm) contributing typical activity values of 700 Bq/kg, 40 Bq/kg and 30 Bq/kg respectively. The activity in coal is only about 10% of that of the shale, and is probably due to the presence of the same sedimentary constituents in lower concentration.

The consistency of these activity levels, at least within a locality in one seam, and more generally from one seam to another, were important factors to be established. Fifty coal and shale samples were collected from one seam over the course of the life of a face covering in all during this period some 1 Km^2 of seam area. Considerable variation in the activity was unfortunately found, though the spectral composition was fairly constant, reflecting constant proportions of the constituent nuclides. It was further discovered, however, that the samples taken from distances deeper into the shale stratum were giving rise to the higher activities. It was found possible to express the activity as a linear function of distance from the coal-rock interface:-

$$x\ (Bq/kg) = 600 + 2h \quad h\ in\ mm$$

With this extension to the input data taken into account, the consistency of samples across the whole are covered by the face panel was within 5%, a figure which equalled the measurement accuracy of the equipment employed.

The model upon which the design of the probe was based consisted of treating the overlying strata as a large plane γ-source and the intervening coal as an absorber whose thickness could be determined by the amount of the attenuation imposed on the γ-flux from the roof. Theoretical predictions of the γ-counts were made for varying coal thickness; initially assuming that only uncollided flux was relevant. The shale activity was approximated by a 6-layer planar sandwich. The results were encouraging, indicating acceptable crystal sizes for a scintillation detector. More detailed calculations using Monte-Carlo methods were subsequently carried out to estimate the total flux present including the collided flux contribution. At the shale surface these indicated

a total count rate of approximately 260 counts/sec for a
75 mm diameter x 75 mm sodium iodide crystal.

Practical field measurements were also carried out to
confirm the theoretical estimates. A special monitor was
constructed and certified for use underground to enable
the γ-flux to be measured at various locations in one
seam and in different seams. This also included
facilities to vary the collimation or entry angle to the
detector. Bearing in mind the consistency obtained in a
detailed survey at one mine, the data was concentrated
upon a few seams in widely separated areas of the country
- Staffordshire, Northumberland, S Wales and Scotland.
Excepting S Wales where special difficulties arose due to
an intervening black, low activity shale, all the results
produced a consistent pattern, as shown in Figure 17.8.

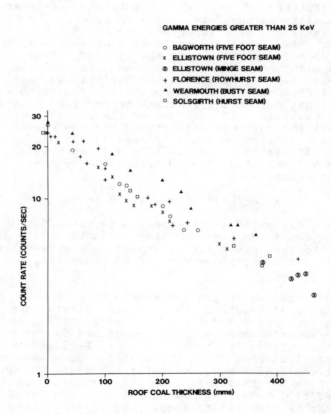

Fig. 17.8   Performance variation in different seams

These certainly confirmed the expectation that a single
design of probe would cover a large proportion of seams.

It has already been noted that sandstones are of much lower activity, roughly 1/3 that of shales; but a sandstone-roof is not common and also presents other mining problems.

Finally, confirmation of the substantial independence of the sensor with 'stand-off' distance from the roof was also obtained. Since the roof area from which radiation is received increases with stand-off distance, it is necessary that there is no other screening feature within the effective window angle of the detector. For example, if the effective roof area within this window is screened by a roof support beam, or includes part of the vertical face, lower levels of radiation will be received. This point must be borne in mind when larger stand-off distances are required. The price of operating with greater collimation i.e. narrower window angles is a lower count-rate.

17.6.3 Design of the Type 801 Probe

Detailed design of the Type 801 sensor was carried out by the Salford Electrical Instrument Co., under the guidance of the MRDE scientists who had been responsible for the initial studies reported above. The choice of detector volume must be a compromise between large crystals and high count rates (to achieve low statistical count-rate errors in times of the order of a few seconds) and small physical size and robustness. The latter are required to present a low interference factor in the operation of mining machines and to ensure reliable operation under severe vibration and shock loading. As noted above in the comments on collimation, the position of the sensor on the machine and the stand-off distance from the roof also affect the proportion of the available flux detected, and hence the crystal volume needed. These factors also vary with the application, that for a shearer machine for example presenting a more demanding solution than for another machine, the in-seam miner, which performs a different function and moves far more slowly.

The main application, for shearer horizon control, led to the selection of a 150 mm x 75 mm diameter crystal of CsI, a crystal which is mechanical stronger than the cheaper NaI alternative. This is operated with the crystal and photomultiplier lying along the length of the probe and radiation entering through the upper surface as may be seen in Figure 17.7.

The lower surface and sides of the probe must be screened from other sources of radiation, especially that which comes from the floor and the waste-area of collapsed overburden strata, since these provide comparable radiation levels to that from the shale layer in the roof. Lead screening of 12 mm thickness along the bottom and sides of the probe provided adequate attenuation to this unwanted radiation, resulting in a background count rate of 32 counts/sec compared with a count of approximately 300 counts/sec at roof level (zero coal thickness). A number of experiments were also necessary to arrive at the effective viewing angle, which is controlled by the height of the side screening. These examined the compromise to be struck between the independence of the coal thickness reading with stand-off distance (favouring a narrow angle to avoid inclusion of roof beams or the face within the field of view) and the count rate magnitude, and associated statistical count rate error for a given time constant (favouring a wide angle).

The high-voltage photomultiplier power supply and count-rate circuits were encapsulated for environmental protection and as part of the design features for intrinsic safety. The circuits provide both digital and smoothed analogue outputs, the latter conforming to standard NCB transducer outputs of 0.4V to 2.0V for the zero to full span coal thickness signal.

A robust mechanical design was obviously essential but in addition, the crystal and photomultiplier being comparatively fragile components, required appropriate vibration and shock resistant mountings. The requirement was to reduce shocks of 40g to values well below 10g which was the quoted withstand-figure for the photomultiplier tube. This was achieved by suspending the complete interior mass using rubber mounts within end cradles, as may be appreciated by examining Figure 17.9. The original prototype was also subjected to a confirmatory field trial. Exposure of the unit to 126 shears mounted on an operational shearer produced no adverse effects.

The complete Type 801 unit, shown in Figure 9 is appreciably smaller (640 mm) and lighter (52 kg) than its predecessor the Type 709 probe; and it has greater range (500 mm) and is substantially independent of mounting distance from the roof (up 400 mm). Its 1983 cost is approximately £8000.

17.7. APPLICATION OF THE NATURAL RADIATION PROBE

The probe is only one component in a complete equipment for horizon control of shearers, so that application is necessarily tied to the introduction of such systems. In fact, design, development and early trials of the sensor during the period 1975-1978 (which followed the investigations of 1972-1975) were paralleled by a complete reappraisal and redesign of the shearer control system as a whole. As noted at the end of 17.5, a new variant of cutting machine, the ranging drum shearer became predominant. In responding to this challenge, the control strategy was radically changed, as also was the technology to implement it. The earlier

Case study 273

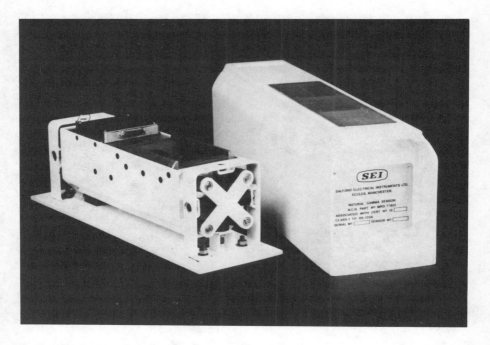

Fig. 17.9   View of type 801 probe

fixed drum machine control system was an analogue system which has been described by Webb (9). The new version developed by Barham and Wolfenden (10, 11) was a digital system based on a low-power CMOS microcomputer.

The new probes found application in three ways

- as much more attractive alternatives to the Type 709 backscatter probe for fixed drum machine systems (using the older analogue control system)

- as elements in the new digital horizon control system for ranging drum shearers (and later also for fixed drum machines)

- as indicators only, providing a visual thickness indication for manual steering action by the operator.

Figure 17.10 illustrates, as an example, its application to a single-ended ranging drum shearer; it may be compared with the earlier Type 709 probe arrangements in Figure 17.5. The lack of intrusive mounting gear is very evident, and this has been a very important element in its rapidly widening application. When used in the third role, i.e. as an aid to manual

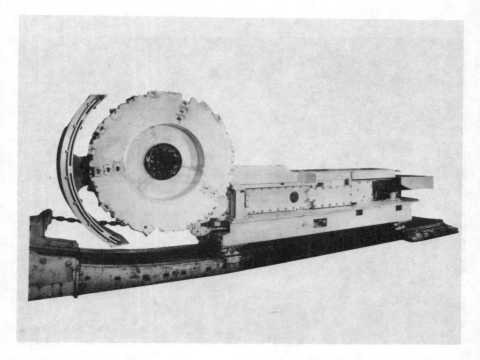

Figure 10   Single-ended ranging drum shearer fitted with type 801 probe

steering, a small display unit is suitably mounted for the operator. As well as a digital coal thickness reading, this provides a simple illuminated pattern of lights for the driver, in the form of a vertical line of light-emitting-diodes. For thicknesses less than the selected nominal value, red lights are progressively lit further up the line; conversely, for excess thickness, green lights are lit below the centre line; while for a correct value a yellow horizontal line is lit.

In all, some 70 natural gamma sensors have been supplied to the National Coal Board, of which about 50% are in current use principally for the indicator role. However, the rate of application is currently affected by the availability of a production version of the microcomputer-based control system equipment. The limit of 6 to 8 experimental installations is now installed or committed; but requirements exist for more than 30 installations when the production equipment has been proved and made available.

Overseas interest has also been marked both in the West and in Eastern block countries. One supplier of machines for high wall extraction, a type of mining from an open hill-side or cliff practised in the USA, has purchased 120 units. A Queen's Award for Technological Merit was awarded jointly to the Mining Research and Development Establishment and Salford Electrical Instruments Limited in 1982, for this development and its application to horizon control of cutting machines.

17.8. POSTSCRIPT - THE EVOLUTION OF A RESEARCH AND DEVELOPMENT PROJECT

The phases of a major research and development project can be summarised, at least from a conceptual viewpoint, by the sequence shown in Table 17.1.

In practice, depending upon the importance of the objective and the resources available, the phases may be elaborated or concertina-ed; and also more importantly feed back loops between various phases are present, only one of which is represented diagrammatically by the dotted line; and new ideas will, one hopes, be welcome at any time!

How far, one may ask, does the experience of the long development of coal thickness measurement devices provide a confirmation or commentary upon this development sequence? The principal comments are

- the requirement, in a field where little experience was available, was itself evolving. While a range of thickness measurement up to 150 mm was thought to be adequate initially, experience showed this to be too low; but of still greater importance (and a counter factor) the sensing device interfered with normal operations and this proved to be a severe limitation. As a general observation, effort to establish the objective, not merely the ideal objective, but with all the compromises and qualifications that inevitably accompany realistic targets, i.e. in as much depth as possible and as early as possible, is likely to be very beneficial.

- the contribution of a number of disciplines to the project is likely to improve several aspects:- a greater realism in the statement of the requirement (with restraints identified and lower limits of tolerable performance recognised); better selection of preferred solutions; and efficient testing and evaluation of prototypes. In this project mining engineers as well as physicists, electronic (and later computer) engineers were involved, with considerable advantage.

- the time-scale for each iteration of development was between 3 and 7 years, depending upon the degree of novelty involved; and it is rash to expect that in a new field less than 2 or 3 iterations will be needed. Particular factors in a mining context are the need at each redesign to include approximately 1 year for certification of electrical equipment; and in a difficult environment, often with production constraints field testing is also a lengthy phase.

- the maintenance of a 'window' for new ideas was very important; the most successful solution did not come from the 'traditional' review of alternatives.

- a conceptual appreciation of the characteristics of alternatives was also vital; the planar source of radiation and its consequences - the positional independence of the detector - were key factors in the decision to investigate this alternative solution. (The practical pressures to improve an existing method rather than embark on a new one with a big risk are considerable!)

These remarks are relevant to a project in which the research element was considerable. In the later stages of product development such as in enhancement, optimising and engineering for production, the weighting of several factors would no doubt be different.

TABLE 17.1   Phases of a research and development
             programme

- statement of broad requirement
    desired performance
    minimum performance
    environmental constraints

- review of existing knowledge
    previous solutions and limitations
    (other countries, other industries)

- selection of potential solutions
    earlier solutions
    new concepts (brainstorming, new technology)

- evaluation of one or more priority solutions
    theoretical/modelling solutions
    problems, limitations, costs
    design specification

- prototype design
    modelling, calculations
    compromises, trade-offs
    electronic, computer, mechanical design:
    bench/rig evaluation: performance, problems,
                          costs, reliability
    prototype evaluation: performance, problems,
                          costs, reliability

- field trials of prototype: performance, problems,
                              costs, reliability

17.10   ACKNOWLEDGEMENTS

The work described has been carried out by many colleagues at MRDE over a long period, many of whom are no longer members of the MRDE; I acknowledge collectively their major contribution to this project. The views expressed however, are those of the author alone. I am also grateful to the Director, P G Tregelles for his kind permission to present this paper.

REFERENCES

1. Symposium on Remote Control of Electrical and Mechanical Equipmen at the Coalface, Ass. of Mining Mech. and Elec. Eng., Harrogate, Nov. 1964.

2. Thomas, V.M., and Pidgeon, B.G., 1964, 'Remote Control of Face Machines : Part 1 Coal Winning Machines'. Proc of Symposium on Remote Control of Elec and Mech Equipment at the Coalface, Ass of Mining Mech and Elec Engineers, Harrogate 61-81.

3. Becque, P.J., and Thomas, V.M., 1963, 'Control Equipment for a Remotely Controlled Mole Miner - the Collins Miner'. Min Engineer 122, (33), 647-666.

4. Terry, N.B., and Seaborne, R.F., 1958, 'The Elastic Properties of Coal, Part 5; An Apparatus for Measuring Elastic Pulse Velocities in Coal Specimens'. J Brit Inst Radio Eng., 18, 371-380.

5. Nosov, G.R., 1965, 'High Frequency Non-contact Measurement of Coal Layer Thickness'. Gornaia Elektromekhanika i Avtomatica No 1.

6. Broussard, P., and Schmidt, W.B., 1981, 'The Longwall Automation Research Project of the U.S. Dept of Energy'. Mining Tech. 63 (726), 138-143.

7. Hartley, D., 1971, 'Automatic Steering of the Shearer Loader at Wolstanton Colliery' Mining Eng 130 (124), 221-236.

8. Wykes, J.S., Adsley, I., Cooper, L.R., and Croke, G.M., 1983, 'Natural -Radiation : A Steering Guide in Coal Seams'. Int. J. Appl. Raidat. Isot. 34 (1) 23-26.

9. Webb, R.E., 1967, 'Automatic Vertical Steering for the Anderton Shearer Loader'. Mining Eng. 126, (80), 531-538.

10. Barham, D.K; and Wolfenden, R., 1979, 'On-line Computer Control of Coal Cutting Machines'. IEE Int. Conf. 'Trends in On-line Control Systems', Sheffield 91-95.

11. Wolfenden, R., 1976. 'Analysis of the Stability of Vertical Steering Systems with particular reference to the Fixed Drum Shearer'. National Coal Board, Mining Research and Development Establishment Report IR 76/41.

TABLE 17.1   Phases of a research and development
             programme

- statement of broad requirement
      desired performance
      minimum performance
      environmental constraints

- review of existing knowledge
      previous solutions and limitations
      (other countries, other industries)

- selection of potential solutions
      earlier solutions
      new concepts (brainstorming, new technology)

- evaluation of one or more priority solutions
      theoretical/modelling solutions
      problems, limitations, costs
      design specification

- prototype design
      modelling, calculations
      compromises, trade-offs
      electronic, computer, mechanical design:
      bench/rig evaluation: performance, problems,
                            costs, reliability
      prototype evaluation: performance, problems,
                            costs, reliability

- field trials of prototype: performance, problems,
                             costs, reliability

Index

Accuracy, 34, 48, 175, 207
Actuators, 155
    "   Choice of, 162
          Electric, 160
          Piston, 158
          Pneumatic diaphragm, 155
          Power sources for, 161
Adsorption chromatography, 128
Analytical instrumentation, automatic, 229
Analytical measurement, 206
Automatic inspection, 191
Automatic titration, 233

Ballistic weighing, 115
Bandwidth, 182
BASIC, 249
Beam balance, 106
Bellows gauge, 95, 120
Bernouilli equation, 26
Bipolar excitation, 14
Black body radiation, 84
Bourdon tube, 95
Bubble pipe, 143
Buoyancy level measurement, 140
Buoyancy weight measurement, 115
Butterfly valve, 153

Capacitance flow transducer, 65, 67
Capacitance pressure transducer, 97
Capacitance probe, 145
Capsuler pressure gauge, 95
Cardiac surgery, 236
Chord type orifice plate, 29
Chromatographic techniques, 211, 224
Classification of errors, 165
Coal cutting machine, 251
Coal mining system, 251
Coal seam thickness measurement,
        Electromagnetic, 256, 259
        Mechanical, 256, 260
        Nucleonic, 256, 257, 261
        Sonic/ultrasonic, 256, 259

Cold junction, 79
Conduction errors, 86
Conductivity flow transducer, 65
Continuous belt weighing, 66
Control valves,
    Rotary, 152
    Sizing of, 154
    Sliding stem, 149
Control valve characteristics,
    Equal percentage, 149
    Linear, 149
    Quick opening, 149
Coriolis meter, 68
Correlation flow measurement, 51, 63, 243
Correlogram, 246
Cross correlator, 65

Dead weight tester, 94
Decision making - inspection, 192, 201
Defect parameters, 202
Differential pressure flowmeter, 24, 241
Differential pressure level measurement, 144
Digital electronics, 236
Dimensional analysis, 123
Discharge coefficient, 25, 31
Displacement level instrument, 140
Doppler flowmeter, 48, 51

Eccentric disc valve, 153
Electric valve actuator, 160
Electrochemical techniques, 211, 220
Electromagnetic flowmeter, 1
Electron spin resonance, 218
Electrostatic flow transducer, 65
emf/temperature relationship, 80
epROM, 249
Equal percentage valve characteristic, 149
Error models, 186
Errors in measurement, 164
    Classification of, 165
    Distribution of, 173
    Influence errors, 166, 169
    Intrinsic errors, 166
    Random errors, 171, 186
    Systematic errors, 171, 185
Expansion thermometer, 76

Fail safe valve, 162
Finite element model, 130
Float actuated level measurement, 138
Flow coefficient, 27
Flow measurement - two phase, 61
Flowmeters,
    Comparison of types, 18
    Cross correlation, 51, 63
    Doppler, 48, 51

Electromagnetic, 1
Orifice plate, 25
Positive displacement, 38
Relative usage of, 1
Turbine, 40
Ultrasonic, 41
Variable area, 52
Vortex shedding, 54, 70
Force balance, 106
Force balance diaphragm, 143
Force measurement, 105
Force transducer, 124
Functional modelling, 120

Gamma radiation probe, 261
Gamma rays, 257
Gamma ray sensor, 267, 269
Gas chromatography, 229, 238
Graphics - instrumentation, 246
Gyroscopic load cell, 116

Hook gauge, 137
Hydraulic load cell, 108
Hydrostatic level measurement, 142

Immersion thermometers, 86
Influence errors, 166, 169
Infrared spectrometer, 214, 218
Inspection - automatic, 191
Interrogation, 192, 194
Intrinsic errors, 166
Ion exchange chromatography, 228
IPTS, 74

Laser sensing, 194
Linearity, 184
Linear valve characteristic, 149
Liquid chromatography, 224
Liquid level measurement, 136
Load cell, 108-117, 125
LVDT, 98, 110, 120

Magnetic flowmeter, 63
Magnetoelastic load cell, 117
Manometer, 92
Mass flow, 61, 66
Mass spectrometry, 218, 238
Mathematical modelling, 119
Measuring system errors, 164
MEDIEM, 121
Microprocessors, 235
Microwave spectrscopy, 218
Models, 119, 193
Modulation, 192, 194

NASCOM 2, 250
Normal distribution, 208

Nuclear magnetic resonance, 173, 218
Nucleonic techniques, 69

Optical inspection, 193
Optical transducer, 65
Orifice plate flow measurement, 25, 241

Pattern recognition, 198
Peak search, 246
Peak tracking, 248
Photomultiplier, 194
Physical models, 121
Piezoelectric load cell, 116
Piezoelectric pressure transducer, 99
Pipeline conveying, 61
Planck radiation, 83
Pneumatic conveying, 62, 240
Pneumatic load cell, 108
Polarity correlation, 244
Polarography, 222, 224
Positioners - valve, 156
Positive displacement flowmeter, 38
Potentiometric pressure transducer, 102
Potentiometry, 222
Pressure gauge, 142
Pressure measurement, 91-104
Pressure tappings for orifice plates, 27, 30, 32
Pressure transducers, 97-103
Pyrometry, 83

Quick opening valves, 149

Radiation, 214
Radiation thermometer, 83
Radio frequency level measurement, 146
Raman spectroscopy, 218
Random errors, 186
Rangeability, 25
Reference junction, 79
Reluctive pressure transducer, 98
Repeatability, 175
Reproducibility, 175
Resistance thermometer, 81
Resistive pressure transducer, 103
Resistivity, 82
Resolution, 184
Rotary control valves, 152
Reynold's number, 29

Sensing - error, 192, 194
Sensitivity analysis, 123
Shear force load cell, 130
Signal detection, 199
Signal delineation, 199
Signal parameterisation, 199
'Sing around', 44
Sinusoidal excitation, 9

Slurry flow, 62
SO_2 scrubber, 22
Spectral radiance, 83
Spectroscopic techniques, 211, 213
Stagnation temperature, 87
Stolz equation, 32
Strain gauge, 100, 112, 114, 125
Systematic errors, 185

Temperature measurement, 73-90
Thermistor, 82
Thermocouples, 65, 77
Torque tube, 141
Transducers, modelling of, 121
Turbine flowmeter, 40, 70
Two phase flow measurement, 61-72

Ultrasonic flowmetering, 41, 65, 66
Unidirectional excitation, 12
U-tube, 93, 144
U/V Spectroscopy, 218

Valves - control, 148
Valve sizing, 154
Variable area flowmeter, 52
Vibrating probe level measurement, 145
Vibrating wire pressure transducer, 102, 109
V-notch ball valve, 152
Voltammetry, 222
Vortex shedding flowmeter, 54, 70

X-ray spectroscopy, 218

£17.00
ML

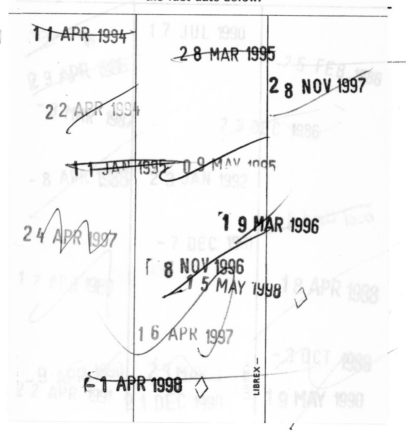

Books are to be returned on or before the last date below.